ON RARE BIRDS

ANITA ALBUS

Translated from the German by

GERALD CHAPPLE

ON
Rare Birds

LYONS PRESS
Guilford, Connecticut
An imprint of Globe Pequot Press

FOR ISABEL & MAXIMILIAN

Georg Flegel, *Still Life with Kingfisher*
(Alcedo ispida), early 1600s

Contents

Translator's Note

T HERE HAVE been a few changes made in transposing *Von seltenen Vögeln* into English: the definitions in the original glossary have been woven into the text; the list of abbreviated references has been transformed into a bibliography, which includes references that are found throughout the book. The list of names and measurements has been dropped, as the hundreds of dialect names found there would make little sense to a non-German-speaking audience. I have added some endnote material in square brackets. All the illustrations that appeared in the original edition have been retained here, and one has been added.

Anita Albus translated George Louis Leclerc, Comte de Buffon's treatise from the French in its entirety, partly to capture the detail, grace, and flavor of his eighteenth-century prose. William Smellie's 1793 English translation, although abridged, has the virtue of more clarity while preserving, for us, the experience of reading the older language. Smellie's translation was also the most popular of the three early English versions of Buffon's text.

The pleasant task of thanking people must begin with the author. Anita Albus read the translated manuscript thoroughly and instructively, enhancing drafts unstintingly with her fine knowledge of English and clearing up translation cruxes. She was ably assisted in Munich by Wolfgang Heuss, whose linguistic skill and ornithological expertise were invaluable. Robert Curry read the manuscript with a keen bird expert's eye, for which I am most grateful. My thanks also go to Rob Sanders and Greystone Books for taking on this splendid project and to Nancy Flight, Lara Kordic, Michael Mundhenk, and Carra Simpson for their professionalism and warm support throughout. Barbara Perlmutter and McMaster University helped out in various ways. I am of course responsible for any errors or inaccuracies that remain and would appreciate any readers' corrections for future printings.

I owe a final debt of gratitude to Nina, my wife, for her constant help in scrutinizing the manuscript about these birds and her companionship in appreciating them and their less rare kin.

GERALD CHAPPLE
Dundas, Ontario
May 2010

PART I

Birds Extinct

CHAPTER 1

The Passenger Pigeon's Eclipse

Thus pass away the generations of men!—thus perish the records of the glory of nations! Yet when every emanation of the human mind has faded—when in the storms of time the monuments of man's creative art are scattered to the dust—an ever new life springs from the bosom of the earth. Unceasingly prolific nature unfolds her germs, regardless though sinful man, ever at war with himself, tramples beneath his foot the ripening fruit!

ALEXANDER VON HUMBOLDT, from
"On the Waterfalls of the Orinoco," *Views of Nature* (1849),
translated by E. C. Otté and Henry G. Bohn[1]

IT WAS a sunny Kentucky day in autumn 1813, when John James Audubon set out for Louisville from his house in Henderson, on the Ohio River. As he was riding briskly over the Barrens toward noon, a few miles beyond Hardensburg, a black cloud of Passenger Pigeons appeared on the horizon. The flocks were flying from northeast to southwest at over sixty miles an hour. The air was soon filled with pigeons, the sky darkening in a flash as if in an eclipse. Bird droppings rained down on Aububon like melting flakes of snow. The unending buzzing of wings began to lull his senses to sleep. To fight this off, he dismounted. He put a hand over his sketchbook while trying to estimate the number of flocks, making a dot with his pencil for each one. One hundred and sixty-three flocks in twenty-one minutes. But that was only a tiny sliver of the huge flock rushing high overhead and stretching from the Ohio River to the huge forests in the far distance. And the end was nowhere in sight.

While Audubon waited for his lunch in an inn at the confluence of the Salt and Ohio rivers, he could see immense multitudes of Passenger Pigeons moving high in the air over the barren countryside. Not one pigeon would land unless some of their millions of fiery red eyes could spy some woods with beech mast or acorns, or fields of wheat or rice for their millions of pitch-black bills. If a falcon tried to seize a bird in the flock, the pigeons quickly closed ranks into a compact mass, generating a roll of thunder with their beating wings. Like a living torrent they plunged down in almost solid masses and "darted forward in undulating and angular lines, descended and swept close over the earth with inconceivable velocity, mounted perpendicularly so as to resemble a vast column, and, when high, were seen wheeling and twisting within their continued lines, which then resembled the coils of a gigantic serpent."[2] Moved by the beauty of the spectacle, this painter of birds observed how one flock after the other would fly into the space where a pigeon had just escaped a falcon's talons, and how, even if no raptor were present, they would form a living river in the air and replicate the angles, curves, and undulations of the attacked flock before them. A single memory bonded millions of pigeons together.

The sun had not set by the time Audubon reached Louisville, fifty-five miles from Hardensburg. There was still no end of the birds in sight. The flock continued passing by for three days. The mass of birds summoned masses of men into the field. "The people were all in arms. The banks of the Ohio were crowded with men and boys, incessantly shooting at the pilgrims, which there flew lower as they passed the river. Multitudes were

Die Wandertaube

Christian Buhle, *The Passenger Pigeon*
(Ectopistes migratorius), 1835

thus destroyed. For a week or more, the population fed on no other flesh than that of Pigeons, and talked of nothing but Pigeons. The atmosphere, during this time, was strongly impregnated with the peculiar odour which emanates from the species."[3]

Audubon estimated afterward that, considering the extent of the flock and the length of time it took to go by, it consisted of 1,115,135,000 Passenger Pigeons. Alexander Wilson, Audubon's rival, came up with an estimate of over two billion. The horrendous mass of pigeons was matched by a horrendous need for food. If the Argus-eyed swarm spotted a food supply, the birds would crowd together to form a feathered serpent, circling and sinking, to seize possession of the land. The plumage of this twisting, serpentine formation, like that of a single pigeon, flashed iridescent purple, green, and gold and shimmered from a reddish color to a slate blue, depending on the vantage point of the viewer, as it maneuvered through the air. Magnificent and fearsome, like the birds of the Apocalypse that were summoned by the angel with "Come and gather yourselves together unto the supper of the great God,"[4] the swarms of Passenger Pigeons ravaged woods and fields.

The worst devastation was wrought in the forests where they roosted and bred. Audubon witnessed a pigeon massacre in their typical woodland habitat of magnificent tall trees and scant underbrush, where the birds had gathered nightly for two weeks:

I arrived . . . nearly two hours before sunset. Few Pigeons were then to be seen, but a great number of persons, with horses and wagons, guns and ammunition, had already established encampments on the borders. Two farmers from the vicinity of Russellville, distant more than a hundred miles, had driven upwards of three hundred hogs to be fattened on the pigeons which were to be slaughtered. Here and there the people employed in plucking and salting what had already been procured, were seen sitting in the midst of large piles of these birds. The dung lay several inches deep, covering the whole extent of the roosting-place . . . Many trees two feet in diameter, I observed, were broken off at no great distance from the ground; and the branches of many of the largest and tallest had given way, as if the forest had been swept by a tornado. Every thing proved to me that the number of birds resorting to this part of the forest must be immense beyond conception. As the period of their arrival approached, their foes anxiously prepared to receive them. Some were furnished with iron-pots containing sulfur, others with torches of pine-knots, many with poles, and the rest with guns. The sun was lost to our view, yet not a Pigeon had arrived. Every thing was ready, and all eyes were gazing on the clear sky, which appeared in glimpses amidst the tall trees. Suddenly there burst forth a general cry of "Here they come!" The noise which they made, though yet distant, reminded me of a hard gale at sea, passing through the rigging of a close-reefed vessel. As the birds arrived and passed over me, I felt a current of air that surprised me. Thousands were soon knocked down by the pole-men. The birds continue[d] to pour in. The fires were lighted, and a magnificent, as well as wonderful and almost terrifying, sight presented itself. The Pigeons, arriving by thousands, alighted everywhere, one above another, until solid masses as long as hogsheads were formed on the branches all round. Here and there the perches gave way under the weight with a crash, and falling to the ground, destroyed hundreds of the birds beneath, forcing down the dense groups with which every stick was loaded. It was a scene of uproar and confusion. I found it quite useless to speak, or even to shout to those persons who were nearest to me. Even the reports of the guns were seldom heard, and I was made aware of the firing only by seeing the shooters reloading.

No one dared venture within the line of devastation. The hogs had been penned up in due time, the picking up of the dead and wounded being left for the next morning's employment. The

Theodore Jasper, *Passenger Pigeon*
(*Ectopistes migratorius*), 1881

Pigeons were constantly coming, and it was past midnight before I perceived a decrease in the number of those that arrived. The uproar continued the whole night; and as I was anxious to know to what distance the sound reached, I sent off a man, accustomed to perambulate the forest, who, returning two hours afterwards, informed me he had heard it distinctly when three miles distant from the spot. Towards the approach of day, the noise in some measure subsided, long before objects were distinguishable, the Pigeons began to move off in a direction quite different from that in which they had arrived the evening before, and at sunrise all that were able to fly had disappeared. The howlings of the wolves now reached our ears, and the foxes, lynxes, cougars, bears, raccoons, opossums and pole-cats were seen sneaking off, whilst eagles and hawks of different species, accompanied by a crowd of vultures, came to supplant them, and enjoy their share of the spoil.

It was then that the authors of all this devastation began their entry amongst the dead, the dying, and the mangled. The Pigeons were picked up and piled in heaps, until each had as many as he could possibly dispose of, when the hogs were let loose to feed on the remainder.[5]

SIMILAR SLAUGHTERS would take place in the deciduous forests, where the Passenger Pigeon had established gigantic breeding colonies. The flesh of older birds was tough and dry, but the nestlings' was tender and unusually fatty. The Native population had long used Passenger Pigeons as a source of food, and white settlers had learned from the Natives to substitute the bird's fat for butter and lard. It was the Aboriginals who prophesied that it would not be long before there were no more pigeons. But no good Christian was about to believe that.[6] Pigeons were suffocated by sulfur fumes, beaten to death with sticks and clubs, caught by the thousands in traps and nets, showered with lead and birdshot, and decimated when trees with hundreds of nests and nestlings were felled.

Technical advances aided and abetted the extirpation: news of nesting sites went by telegraph; heaps of pigeons were transported by rail. It was said that three billion Passenger Pigeons in eastern North America would migrate in search of food all the way from Nova Scotia to the Gulf of Mexico. By the end of the 1880s, *Ectopistes migratorius* had become rare compared with the millions upon millions that had once plunged whole expanses of land into darkness during migration. There may have been a few thousand still left at the time. Scattered flocks were sighted over feeding areas in Michigan, Wisconsin, Indiana, and Nebraska. They were still being hunted, but pursuing them with nets, traps, poles, and sulfur was no longer worth the effort. If the Passenger Pigeon had once seemed to the white settlers like three of the nine plagues of Egypt— locusts, hail, and darkness—rolled into one, now the prophecy of the Natives was to be fulfilled. Nature completed the business of destruction wrought by human masses. The superorganism of the giant flock was cut to shreds. The birds, reduced to a few sparse flocks and driven by herd instinct, were now no match for their natural predators. Thunderstorms and hail in the breeding season, forest fires, and viral epidemics took care of the remainder.

The last *Ectopistes migratorius* to survive in the wild was a female. The fourteen-year-old son of an Ohio farmer shot her on March 24, 1900. The stuffed bird can be seen today in a museum in Columbus.[7] The few birds kept in captivity in zoos did not breed. The last survivors of the species could be seen in 1909 in the Cincinnati Zoo: two males and a female. A year later the female was all by herself. She was named Martha, not after Lazarus's sister, but after George Washington's young widow. Martha's late male companion had been called George, after the president. Nobody can warm up to a million birds, but every heart went out to lonesome Martha. The last of her species, she was *the* attraction for hordes of visitors. She died on September 1, 1914; her body was frozen in a large block of ice and brought to Washington to be carefully examined in the Smithsonian. Afterward, her stuffed skin was put on display.

John James Audubon, *Pair of Passenger Pigeons*
(*Ectopistes migratorius*), pl. LXII, 1827–38

Stuffed specimens also served as models for the pair of pigeons in Audubon's idyllic painting. This highly gifted autodidact would sketch from life, and after shooting the birds he would prepare them in the attitude in which he had observed them so that he could reproduce their beauty of movement and plumage at his leisure.[8] He vividly portrayed how the male's courtship display ultimately led to the pair's billing, where male and female would take turns stuffing the contents of their crops into the other's mouth, as we see in his illustration. The male provides the female with food during the breeding season. The tenderness and dedication shown by the males to their mates were, for Audubon, "in the highest degree striking."[9]

As the Passenger Pigeon was becoming rarer, the "man of the forests and savannahs" had long passed away.[10] He could not have known that his painting—with its dead leaves and lichen-laden branch on which a male pigeon, fanning his tail, is letting the female put food in his crop—would one day recall to us the last pair of Passenger Pigeons on this earth: George and Martha, billing. Their brother was never resurrected.

CHAPTER 2

The Parakeet and Its Practices

A SECOND, VERY different pair of lonely birds just happened to be in the Cincinnati Zoo during Martha's time there. They, too, were the last of their species. *Conuropsis carolinensis*, the Carolina Parakeet, shared the Passenger Pigeon's fate. They, too, would appear in masses of closely knit flocks; they, too, as intelligent as they were, could not help but flock together, which favored the war of extermination against them. Their social behavior was very elaborate, as is that of all parrots. If gunshots spooked them in their feeding area, they would always return to their wounded and dying companions, swooping and screaming over them—and be felled *en masse*.

Audubon had already bemoaned the fact that flocks of the sole North American parrot species were becoming rarer and rarer in the continent's cypress swamps and deciduous forests. His painting shows a band of parrots—a green-headed juvenile among them—going after cocklebur seed pods, their favorite food. They were welcomed by farmers for consuming the weed seeds, but the birds' nourishment was not to be restricted to the seed in the prickly weed capsules and the seed fruits of the forest. Other seeds were to be savored. They could tear unripe apples and pears off the tree with their dexterous "foot-hands." Their powerful

John James Audubon, *Carolina Parakeets*
(Conuropsis carolinensis), pl. XXVI, 1827–38

bills quickly pecked through the small fruit's hard flesh and laid the core bare. The soft, milky seeds proved to yield little. Entire orchards were required to still the parrots' hunger. Bright flocks of them would also descend onto shocks of wheat in the fields, pull out the stalks, and "destroy twice as much of the grain as would suffice to satisfy their hunger."[1] The colorful birds would often completely blanket the sheaves, so that it seemed, in Audubon's eyes, "as if a brilliantly coloured carpet had been thrown over them."[2]

The Carolina Parakeet was particularly attracted to large cypresses and sycamores and would roost in their hollows, "thirty or forty, and sometimes more, entering at the same hole. Here they cling close to the sides of the tree, holding fast by the claws and also by the bills. They appear to be fond of sleep, and often retire to their holes during the day, probably to take their regular siesta."[3]

Woodpeckerlike calls would accompany their particularly rapid and graceful flight, which resembled the Passenger Pigeon's spirals. A gyrating flock of parakeets would flash the iridescent colors of its spectacular plumage—sometimes the upper feathers, sometimes the underside— at the observer. The green-yellow-red birds would also delight the eye as tree decorations. A German settler in Missouri recorded in his 1877 autobiography the "particularly attractive" sight of "a flock of several hundred [parakeets] settled on a big sycamore, when the bright green color of the birds was in such marked contrast with the white bark of the trees, and when the sun shone brightly... the many yellow heads looked like so many candles." This reminded him of a kind of Christmas tree found in Germany's poorer families: a young birch would be put in a bucket of water some weeks before Christmas and, in "the warm room, soon began to produce delicate leaves [and was then] decorated with gilded and silver nuts and with apples and candles."[4]

The Carolina Parakeet's territory extended as far north as the Great Lakes. Wilson marveled at a swarm's shrill calls as it flew along the banks of the Ohio in February. The parakeets native to Louisiana in the South were different from the northern species in that their green feathers were tinged with blue.

"Polly" was the name Wilson gave to an injured female Carolina Parakeet that he found and wrapped in a silk handkerchief and took with him before putting her into a cage on someone's front porch in Natchez. Her call caught the attention of passing flocks:

> Such is the attachment they have for each other. Numerous parties frequently alighted on the trees immediately above, keeping up a constant conversation with the prisoner. One of these I wounded slightly in the wing, and the pleasure Poll expressed on meeting with this new companion was really amusing. She crept close up to it as it hung on the side of the cage; chattered to it in a low tone of voice, as if sympathizing in its misfortune; scratched about its head and neck with her bill; and both at night nestled as close as possible to each other, sometimes Poll's head being thrust among the plumage of the other. On the death of this companion, she appeared restless and inconsolable for several days. On reaching New Orleans, I placed a looking-glass beside the place where she usually sat, and the instant she perceived her image, all her former fondness seemed to return, so that she could scarcely absent herself from it a moment. It was evident that she was completely deceived. Always when evening drew on, and often during the day, she laid her head close to that of the image in the glass and began to doze with great composure and satisfaction.[5]

Before the green birds died out toward the end of the nineteenth century, they could be bought cheaply on the pet market. Eugène Rey, a grocer and parakeet fan, described the mischievous character of *Conuropsis carolinensis*:

For many years I have been keeping parrots and, hanging in my window, some Carolina Parakeets. In spite of their not exactly pleasant screams and in spite of their insatiable appetite, they have won my affection through other, very likeable qualities to such an extent that I have never been able to bring myself to get rid of them. These birds had taken but a brief while to get accustomed to me so that, for example, they would readily fly to my hand or head when I held out a walnut, which they take particular delight in eating. If I took the nut in such a way that it was completely covered, the birds would stay calmly at their observation post. But if I cracked the nut in my hand without letting them see it, the crunching sound would summon them to me at once. When I put them into a cage later on, they gave me another opportunity to become acquainted with their great talent. One of their most common bad habits was to immediately upset the water bowl after their thirst was slaked, or throw it out of the cage door onto the ground, while displaying their glee most unambiguously when their mischievousness had the desired success, that is, when the bowl broke. Every attempt to attach the bowl or keep the cage door closed failed because the birds, owing to their undaunted efforts learned all too quickly how to remove these barriers. Since I could not get anywhere with these devices, I chose another tack by sprinkling the birds with water every time I caught them being naughty. It afforded an indescribably comical sight, when they, working jointly, would surreptitiously open the sliding door: the bird below would use his bill as a lever and the other, hanging from the top of the cage, would hold onto the door with all his might until his partner had lifted it up another little bit. When the opening was large enough, after a short time, to allow the lower operator to get out, he would then very carefully pull the water bowl toward him, which would, if I did not intervene quickly, meet the same fate as had so many of its predecessors.[6]

La Perruche à tête jaune. Pl. 33.

Jacques Barraband, *The Carolina Parakeet*
(*Conuropsis carolinensis*), 1801–5

To the rascally Carolina Parakeets, Eugène Rey must have seemed like the comical bird. What tricks they had to pull on him to get sprinkled with water, which their feathers so needed! All parrots love the rain. Modern zoos even have special sprinkler systems in their aviaries to provide comfort and freshen up parakeets' feathers.

It is not known where and when the last Carolina Parakeets in the wild died. The little rascals that still remained in captivity were granted but a short reprieve. The last Carolina Parakeet in the Cincinnati Zoo went by the name of Incas. After his mate, Lady Jane, died in late summer 1917, he pined for months before finally passing away himself amid his keepers on a February evening in 1918. Incas's body was to be autopsied and stuffed in Washington, as Martha's and Lady Jane's had been. But the block of ice with Incas inside never arrived at the Smithsonian. The riddle of how and where it melted has never been solved.

CHAPTER 3

The Ways of the Great Auk

L ONG BEFORE penguins were discovered, another bird species bore
that name, but the last surviving pair of this species could never
have been put in a zoo. Their world extended from the peaks
and troughs of ocean waves down into the depths of the shellfish banks
around Eldey, a small, rocky island off Cape Reykjanes on the southwest-
ern tip of Iceland. Like all birds of their primeval genus, the pair shunned
terra firma. They would come ashore on the rocks just to take a breather
after a fierce fish hunt and to oil their feathers and preen. Only the business of breeding called for a longer stay on the "sack of flour," the term
bird catchers used for the oval island rising steeply from the sea like a
stone sack and constantly bedecked with white sea birds resembling a
dusting of flour.

The cliffs slant upward on one side of the flour sack before running to
a considerable height as a steep upper wall. There it was that two flightless birds labored up the cliffs in May 1844. Protected by the rock roof of
the steep wall, they were soon taking turns incubating their sole egg. On
the morning of June 3, while the female was sitting on the egg, they made
a dreadful discovery. Three men were coming up the slope right at them.
Seabird flour flew off in all directions as the flightless couple in their

John Gerrard Keulemans, *Pair of Great Auks (Pinguinus impennis)*,
with the Island of Eldey in the background; the
"flour" from the sea birds is not visible at this distance, 1897–1905

black, feathered tuxedos and white shirts saw a single hope for salvation with their white-bespectacled eyes: the sea.

Before the Great Auk was recognized as *Pinguinus impennis,* it was also called *Plautus impennis,* "non-flying flatfoot." Its Icelandic name, however, is *geirfugl,* "spear-billed bird," because it is swift as an arrow when in its element. Masses of Great Auks used to shoot off the cliffs—on *Geirfuglasker* ("Great Auk Rock," or "The Island of the Great Auks")—into the crystal-clear North Sea waves. Alfred Edmund Brehm said that a Great Auk out of water was "an unfortunate creature tied to one spot," whose movements could not really be called "walking"; even a snail was better on its feet, he added.[1]

Hoping to save themselves by jumping into the surf from a lower cliff, the pair of birds waddled desperately, trying to speed up their flight—heads sticking out and stubby wings rowing the air. People also find it hard to run on cliffs. One of the men grabbed one auk just as it was about to take off into the sea. Two men cornered and caught the second one among the rocks.

A strong wind came up; the surf rose. One of the Icelanders hurried back to the breeding site, only to find that the precious egg had cracked on a block of lava. In a rage, he threw it over the cliff.

The helpless birds were quickly strangled. An agent in Reykjavik, acting on behalf of various museums and collectors, was already waiting for the booty. The egg, if intact, would have fetched as much as one of the skins, which the Museum of Copenhagen ended up buying. Before long, when the bird was extinct, an egg would be worth six thousand crowns, and even a damaged one could garner three or four thousand. By the time almost no collector was willing to part from his rare spinning top, auction prices had climbed to twenty thousand crowns.[2]

No two Great Auk eggs are alike. Their spinning top–like shape prevented them from wobbling off cliffs. They are the largest of all the spotted eggs laid by European birds. Their whitish, yellowish, or greenish shells are thick, large-pored, and dull. They are painted from pointed tip to flat tip with gray, grayish-brown, brown, and black streaks, with loops, squiggles, serpentine lines, meanders, and dots usually on a yellowish ground, as if applied by the brush of a sleepwalking Chinese calligrapher. Every egg has different "lettering."

The eggs had been collected for centuries because they kept well and were delicious. But as the native populations of Iceland, Newfoundland, and Greenland found, not only were the Great Auk's eggs manna from heaven, but so were all parts of the "northern penguin." Its meat was salted down for the long winters; its fat was rendered into oil; parts of its dried skin were used for fuel, the breastbone for making the best fishhooks, the feathers for decoration, the down for warmth. As long as these small populations kept to themselves, the large auk colonies on the skerries off the coast were not affected. It was only through the mass exploitation by trading companies, which cared not a fig about the ratio of deaths to births, that the "northern penguin" was transformed into a rare bird in the eighteenth century.

It didn't take long for the splinter colonies of birds that were continually hunted for food to be extirpated. Small flocks of Great Auk regrouped on remote cliffs. Nature was little involved in its destruction—but that little bit was spectacular: an underwater volcanic eruption swallowed Geirfuglasker, down into the ocean. And only then did it dawn on museum directors that they had neglected to procure skins, skeletons, and eggs for their collections. So the remaining forty-eight auks that had taken refuge on the "flour sack" did not stay undetected for long. A band of bird catchers, working for that same Reykjavik agent, caught them along with their eggs on Eldey during the breeding seasons of 1830 and 1831. Nonbreeding immatures that had not shuffled onto land fled to the sea. There could not have been many. The number of years it took for a Great Auk to breed is not known. The Razorbill (*Alca torda*), the only one of the genus's twenty-two species alive today that the Great Auk is related to, needs three summers. Had the last pair of Great Auks escaped in 1831, they would have frolicked in the waves for thirteen more years;

Great Auk Egg, photograph, n.d.

year after year in the breeding season, the breeding adults would have climbed up the crooked cliffs, incubated their sole egg on the bare rock, and raised a young bird. While it is unlikely that all of their progeny would have survived, it is equally improbable that none would have. Some bird or other would have to have evaded its parents' killers. Perhaps it was granted to the unknown last *Alca impennis* specimen to perish in its element.

John Gerrard Keulemans, *Great Auk Male in Breeding Plumage with Nestling (Alca impennis)*, 1897–1905

1—10 Alca torda L, Tord-Alk.

Great Auk Eggs, in Johann Andreas Naumann,
Naturgeschichte der Vögel Mitteleuropas, photograph, 1897–1905

Edward Lear, *Great Auk*, 1832–37

The Great Auk's disappearance made it popular. In England, people would have a *Great Auk* cigarette with their old ginger wine called *Great Auk's Head*, and the Great Auk could be found among the protagonists in novels and short stories. Symington Grieve's sad tale of the bird's extermination appeared in Edinburgh in 1885.[3]

Evidence from bones shows that the huge Great Auk, weighing in at twelve pounds and measuring almost thirty-four inches long, ranged in prehistoric times along the European coast, from Russia to Spain, from Northern to Southern Italy, from Scotland to Gibraltar. Of the seventy-eight extant skins and stuffed birds, only two are in winter (basic) plumage; all the others are in breeding (alternate) plumage.[4] The nestling's dark gray downy coat has been documented, but the color of the bird's subsequent downy coat and juvenile feathering is unknown, as is the sequence of molts. The Great Auk's "world" went down with the bird. We know next to nothing about its life, its courtship, its mating, its breeding behavior, the nestlings' development, the length of the breeding or the hatching period. Razorbill chicks, which resemble Great Auk chicks very closely but are only half their size, take up to six days to hatch. The Great Auk's calls, the young bird's begging sounds, its rearing, its territorial defense, its threatening, strutting, and deferential positions, its diving maneuvers and feeding habits, its mating rituals, its group behavior—everything that characterizes the Great Auk will remain a mystery forever.

Only by observing other species of alcids, which belong to the order that includes plovers, sandpipers, gulls, and allies (*Charadriiformes*), can we hazard a guess about how the Great Auk might have resembled them. It would have been monogamous, like all members of this family, including the murres, puffins, guillemots, etc.; its young would have hatched, like all alcids, with a downy coat and open eyes, and it would presumably have been a combination of nidifugous and precocial birds, like Razorbill and Common Murre chicks. Maybe Great Auk parents were able to "read" the calligraphy on their eggs. Experiments with Common Murres, which lay spinning top–shaped, different-sized eggs with wildly different designs, have demonstrated that breeding pairs always pick out their own eggs from among scores of others. They then take the eggs back to their rocky nesting site, cradling them between their feet and belly. From then on, they deem any "reading" superfluous. If their egg is switched with another, or even with a stone, they simply go ahead and incubate it. Nature has prepared them for the chance that an egg may roll a short distance against a neighbor's, but they are not programmed against egg switching by human hand. Otherwise they, like all alcids, are better armed against *Homo sapiens* than the Great Auk was. Their wings not only speed up their underwater flight but enable them to take to the air. Man is their worst enemy, now more than ever, but man only appears as such indirectly. Most alcid species have long been under protection in almost all countries. Yet while they may be protected from hunters and egg collectors, they cannot fly away from their greatest dangers. They are helpless, at the mercy of the petroleum plague, of the nylon nets of fishing fleets that destroy hundreds of thousands of murres annually, and of the poisoning of their prey by toxins—just as the Great Auk once was at the mercy of henchmen dispatched by trading companies and museums.

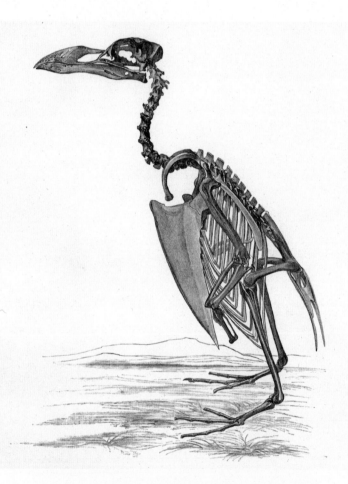

Great Auk Skeleton, in Johann Andreas Naumann,
Naturgeschichte der Vögel Mitteleuropas, 1897–1905

CHAPTER 4

The Loneliness of the Macaw

N OTHING AT ALL survives of the red-shimmering Guadeloupe Violet Macaw, *Anodorhynchus purpurascens*—not a skin, not a skeleton, not a feather. Reports from missionaries and scientific expeditions bear witness to its existence on Guadeloupe in the seventeenth and eighteenth centuries. But the most elaborate description of its plumage is from the sixteenth century and indicates that the macaw was tamed by the Tupinamba, a group of indigenous people residing on the Atlantic coast of Brazil. We owe this description to the French Calvinist Jean de Léry, who lived among the Tupi, as they are sometimes called, from March 1557 to January 1558.

Léry writes in his *Histoire d'un voyage faict en la terre du Brésil:*[1] "[The bird,] which the savages call *arat*, has wing and tail feathers about a foot and a half long, one half of each feather as red as fine scarlet, and the other half a sparkling sky blue (the colors are divided from each other along the quill), with all the rest of the body the color of lapis lazuli; when this bird is in the sunlight, where it is ordinarily to be seen, no eye can weary of gazing upon it."[2] He says the parrot was larger than a raven,

MACROCERCUS ARARAUNA.

Blue & Yellow Macaw.

Edward Lear, *Blue-and-yellow Macaw (Ara ararauna)*, 1832

MACROCERCUS ARACANGA.

Red and Yellow Macaw.

Edward Lear, *Scarlet Macaw (Ara macao)*, 1832

about the size of a blue-and-yellow *Canindé,* the beauty of which the Tupi often praised in song.[3] Léry compares the *Canindé*'s back, wing, and tail plumage to violet-blue damask and its neck and belly to cloth of gold, adding that there is no more wondrous plumage the whole world over than that of these two macaws. Moreover, "although these birds are not domestic, they are more often to be found in the tall trees in the middle of the villages than in the woods, and our Tupinamba pluck them carefully three or four times a year... and use their beautiful feathers to make fine robes, headdresses, bracelets, ornaments for their wooden swords, and other adornments for their bodies."[4]

Claude Lévi-Strauss visited the indigenous people of Brazil almost four hundred years after Léry. In the Bororo village of Kejara, on the Rio Vermelho, he tripped one morning over some "pathetic-looking birds: these were the domesticated macaws which the Indians kept in the village so as to be able to pluck them alive and thus obtain the feathers needed for head-dresses. Stripped of their plumage and unable to fly, the birds looked like chickens ready for the spit and afflicted with particularly enormous beaks since plucking had reduced their body size by half. Other macaws, whose feathers had regrown, were solemnly perched on the roofs like heraldic emblems, enameled gules and azur."[5]

A central motif of many Brazilian indigenous myths is the theft of *Ara* macaw nestlings from breeding sites in rocky niches or tree hollows. The parrot looms large in the Natives' thinking, and not only on account of their beautiful feathers. The Bororo believe "in a complicated system of transmigration of souls: the latter are thought to become embodied for a time in the macaws."[6]

The macaws that the Bororo took from the nest and raised uncaged had no reason to fly off when their feathers had grown in after an induced molt. Imprinted on man and regarded by him as a kindred soul, they were as trusting as the animals in the Garden of Eden.

Long gone are the days of that soulful community of man and macaw:

The Bororo... are today being consumed by alcoholism and disease and are progressively losing their language. It is in missionary schools (which, by a curious reversal, have become the conservators of a culture they had in the first place worked at suppressing and not without success) that Bororo youths are being taught about their myths and their ceremonies. But, for fear that they might damage the feather diadems, masterpieces of traditional art, the missionaries are keeping these objects locked up, entrusting the Indians with them only on strictly necessary occasions. They would be increasingly difficult to replace since the macaws, parrots and other brightly colored birds are also disappearing...[7]

Anodorhynchus purpurascens probably died out in the eighteenth century, and the pastel-toned Glaucous Macaw, *Anodorhynchus glaucus*, one of the four remaining blue macaws, has been missing since the end of the nineteenth century. Eighteenth-century explorers discovered this green and blue, long-tailed bird in the blue-green Yatay palm bottomland forests of the great Paraguay, Paraná, and Uruguay rivers. A Spanish priest the Jesuits had sent as a missionary to the Guarani Indians wrote a manuscript describing the behavior of a tame *Guaa obi*, as the Guarani called the Glaucous Macaw. If a missionary from another station came to the village, he could count on a visit to his quarters from the large *Guaa obi*.[8] If the door was locked, the macaw would climb onto the latch and make a knocking sound with its powerful gray-black beak. Before the door could be opened from inside, the bird would usually have unlocked it by itself. Then it would climb up on the missionary's chair, call *guaa* several times, and move its head most adorably until someone spoke to it as if thanking him for his visit and attentiveness. Afterward, it would climb down and take a satisfied stroll in the yard.[9]

The Yatay palms and their nuts, on which the Glaucous Macaw specialized, disappeared when the fertile bottomlands were cultivated.

Gustav Mützel, *Macaws:* 1. Glaucous Macaw (*Anodorhynchus glaucus*);
2. Hyacinth Macaw (*Anodorhynchus hyacinthinus*); 3. Chestnut-fronted
Macaw (*Ara severa*); 4. Red-and-green Macaw (*Ara chloroptera*);
5. Scarlet Macaw (*Ara macao*); 6. Blue-winged (or Illiger's) Macaw
(*Ara maracana*); 7. Golden-collared Macaw (*Primolius auricollis*), 1878–83

Only a few of these suddenly rare birds found their way into zoos. The last Glaucous Macaw in captivity apparently died in 1905 in the Jardin d'Acclimatation in Paris.

The cultivation of palm savannahs, the draining of wetlands, and the cutting of large trees with nest holes were other threats to *Anodorhynchus hyacinthinus*, the Hyacinth Macaw, a native bird that ranged from the Brazilian northeast to the southeast interior. But an even greater problem for these beautiful birds than habitat destruction was the greed of their aficionados. The Hyacinth Macaw is thirty-nine inches long from head to tail, the longest parrot of all. It is one of the greatest objects of desire for wealthy aviary owners and zoos because of its brilliant deep ultramarine-blue plumage (a touch lighter on its unusually large head) and its playful, amicable, loyal, and teachable nature. The close pair bonding of these birds was first observed in 1990 between a Hyacinth Macaw and its deceased mate in the Walsrode Bird Park in Germany. For six hours the male ran around the dead female lamenting loudly, then clucking softly, until the dead bird was finally torn away from him in spite of his desperate defense of it. A trampled trail in the aviary sand bore witness to his funeral march.[10]

There have long been more Hyacinth Macaws living in captivity than living in the wild. Stealing nestlings is a lucrative business. The rarer the bird, the higher the profit. The poorest of the poor can have no scruples. If you owned little more than a small hut but could now drive a brand-new ATV to go catch birds, then you wouldn't have any qualms about it. In the 1980s alone, ten thousand Hyacinth Macaws were snatched from the wild. At the end of 1987, after the population had dwindled to five thousand, an international ban on dealing in Hyacinth Macaws was instituted. This made the bird all the more desirable. Soon the still precious, giant, ultramarine blue macaw was worth its weight in gold, just as the pigment derived from lapis lazuli had once been. Conservationists campaigned for the Hyacinth Macaw. They were led by Tony Silva, an American from Illinois who had made a name for himself by breeding and writing books on threatened parrots. At the height of his career, he had managed Loro Parque on Tenerife, which housed the world's largest and most spectacular collection of parrots.[11]

Silva published the story of a Paraguayan wildlife trafficker who had obtained three hundred Hyacinth Macaw nestlings in 1972, three of which survived. He also reported on a Brazilian bird catcher who, with a single accomplice, caught two to three hundred Hyacinth Macaws in a span of barely three years. Since the birds do not breed annually and, when they do breed, rarely raise more than one young, Silva had doubts that there would be any Hyacinth Macaws left in the twenty-first century.

Meanwhile, bird trappers and dealers pursued their increasingly profitable illegal business unhampered. Groups of about fifty to eighty Hyacinth Macaws valued at fifteen thousand dollars apiece, along with other rare parrots and wild animals, were taken by ship to the United States with forged source documents. In a motel room in a Chicago suburb, a veterinarian would anesthetize the parrots and open them up with a scalpel—the quickest way to determine their sex.[12]

One person felt so secure in his nature-conservationist skin that he didn't shy away from advertising threatened parrots for sale in *American Cage-Bird Magazine*. It took four years for authorities in the U.S. Fish and Wildlife Service's Special Operations Branch to carry out its investigations under the code name "Operation Renegade." Surveillance reports, wiretaps, and exposed smuggling deals convicted Tony Silva and his accessories. His seven-year sentence began in 1996; his mother got only two years.

They still exist, these gentle, friendly giants of the parrot world, on the border of Amazonia, in the southern parts of Maranhão and Piauí, and in the Pantanal. They still fly in pairs, *en famille*, or in small groups in the morning, and on bright, moonlit nights they crack open nuts in the palm forests. It can still happen that upon leaving their hollows in trees or rock niches they get caught in a trapper's net. The rarer they become, the greater the reward for their capture.

Even rarer than the severely threatened Hyacinth Macaw is the smaller Indigo Macaw, or Lear's Macaw (*Anordorhynchus leari*), which is in danger of becoming extinct.[13] The impassioned ornithologist, Charles Lucien Bonaparte, Napoleon's nephew, published the first definitive description of the Lear's Macaw in 1856 and named the species in honor of the bird painter and nonsense poet Edward Lear. The creator of the limerick had painted the still-unclassified ultramarine-blue macaw with its greenish-blue head and nape for his *Illustrations of the Family of Psittacidae, or Parrots*, in the belief that he had a Hyacinth Macaw in front of him.[14]

How much a Lear's Macaw could develop a sense for nonsense is evident in a story from an English clergyman who thought he was fortunate to own several of these extremely rare birds. His favorite macaw would amuse him with a handkerchief dance: "[A handkerchief] is thrown right on to his head, covering it completely up, upon which he proceeds to dance up and down with great glee, and at last either dances it off, or else pulls it off with one claw, holding it, and then laughing loudly and in a most human way: waiting for the fun to be commenced all over again."[15]

The Reverend Astley didn't know where his Indigo Macaws came from when he wrote about his clownish birds in *The Avicultural Magazine*, the journal of the Avicultural Society. There were a few Lear's Macaws in the London, Berlin, and Hamburg zoos, but no ornithologist had ever seen the bird in the wild in South America. Was it even a separate species? A Dutch zoologist speculated in 1965 that Lear's Macaw was a cross between the Hyacinth and the Glaucous Macaw. Helmut Sick, an ornithologist who lived in Brazil from 1939 to 1991, solved the riddle of how a large blue bird with a wingspan greater than forty inches could go undetected for such a long time. After many grueling expeditions in the sixties and seventies, he set out in 1978 for one of the remotest and most inhospitable areas of Brazil. Exhausted by the rigors of the journey through the endless, almost impassable thorn and cactus of the skeletonlike woodlands in northeastern Bahia, Sick and his team reached, in midsummer, the narrow gorges and the inaccessible sandstone cliffs of the harshest region of the harsh caatinga landscape:[16] the Raso da Catarina.

A hunter had shot and eaten a Lear's Macaw there a few months before. The beautiful feathers he'd saved assured Sick that this time he was on the right track. At dawn on the last day of 1978, he saw for the first time Lear's Macaws flying high over the cliffs toward their palm-nut feeding grounds. Ten days later, the ornithologist spent a most memorable birthday in the Lear's Macaw gorge, seeing the riddle solved before his very eyes through his telescope: at least fifteen Lear's Macaws were flying into the red sandstone caves; one pair stayed for a while in a crevice surrounded by a thick cloud of insects, occasionally scratching their greenish-blue heads.

Ornithologists after Sick were to discover more flocks of Lear's Macaws in other Raso da Catarina gorges. But the survival prospects for the estimated 246 remaining birds are certainly not rosy. One Lear's Macaw consumes up to 350 nuts of the Licury palm (*Syagrus coronata*) a day, but the palms are thinning out, and fewer and fewer pairs of birds have any breeding success. As if that weren't enough, these rare birds are still being trapped for the black market. Smuggling macaws from the wretchedly impoverished caatinga to the world's wealthy countries makes trappers, middlemen, and dealers a fortune commensurate with their respective situations in life.

Since time immemorial, the giant caatinga region by the Rio São Francisco has been dominated by want. The zoologist Johann Baptist Ritter von Spix, who spent almost four years traveling through Brazil with the young botanist Carl Friedrich Philipp von Martius, explored northeastern Bahia and described, in a travelogue written jointly with Martius, the "incredible impoverishment" among "the greatest share of the population."[17] Martius captured the ghostly skeleton-forest landscape in a striking "picture of vegetation" in the first volume of his *Flora Brasiliensis:*

MACROCERCUS HYACINTHINUS.

Hyacinthine Macaw.

Edward Lear, *Lear's Macaw (Anodorhynchus leari),* 1832

It is with dismay and horror that the traveller penetrates this landscape in the dry season. As far as he can see, leafless stems, motionless, unfanned by breezes, stand rigidly around him; no green leaf, no juicy fruit, no fresh blade of grass on the burning barren ground; only strangely shaped cereus stems which rise like monstrous candelabra ... appear to maintain some trace of a fleeting life. But when rain suddenly loosens the fetters of this plant realm, there arises, as if by a stroke of magic, a new world. On the many branches of the stalks leaves of soft green shoot forth, countless kinds of strange flowers make their appearance, the skeletons of the menacing thorn barriers [i.e., hedges] and creepers are clothed in fresh foliage.[18]

The path from Bahia into this desolate world led Martius and Spix past Buriti palm (*Mauritia vinifera* Martius) woodlands; in the palm-tree crowns of waving fanlike leaves they observed numerous "steel-blue *Araras*": "the dwellers in the sublime treetops circled us in pairs and their croaking screeches resounded through the peaceful scenery."[19]

The two explorers made the 620-mile journey through the Brazilian interior mostly on foot. Weakened by tropical fever and hardship, they returned to Munich in 1820, where they were showered with honors and knighted by the king. In his two-volume work on Brazilian birds, Spix named the Hyacinth Macaw *Anodorhynchus maximiliani augusti* in honor of the king who had commissioned him and Martius to go to South America.[20] The new generic name, *Anodorhynchus*, was accepted by scientists, but the species name *hyacinthinus* given by Latham in 1790 was retained.

The "sagacious man of integrity"[21] Ritter von Spix never recovered from the strain of his trip and tropical diseases. After his death in 1826, his assistant, Johann Wagler, was made director of the zoological museum. Wagler discovered that the skin of a small Indigo Macaw that

Carl Friedrich Philipp von Martius, *Caatinga*, 1840

Spix had shot and labeled *Arara hyacinthinus* was not a Glaucous Macaw, as Spix had thought. "Sociable, although extremely rare, characterized in the Rio São Francisco flood plain by a thin call, called 'Guacamayo bleu' in Azara (p. 53)[22] and confused with *Anodorhynchus maximiliani augusti* by Sonnini and surely also by Latham." This is what Wagler read in *Avium species novae* about the bird that shimmered in different tones of blue, was only twenty-two inches long, and that he now recognized as a unique separate species with bright, gray-blue head feathers, dark facial skin, and a dainty bill. In his 1832 *Monographie der Papageien*, Wagler named the new Blue Macaw *Sittace spixii* after its discoverer. But there are many parakeets. Bonaparte, in 1854, gave Spix's Macaw the generic name *Cyanopsitta*, "blue parakeet," that was to be its name from then on.

The skin of the first *Cyanopsitta spixii* known to ornithology can be seen today in the Bavarian State Collection for Zoology in Munich. The last Spix's Macaw living in the gallery forest at the confluence of the Rio São Francisco and a tributary disappeared without a trace in October 2000. As the rarest of all rare birds, the Spix's Macaw, though protected internationally since 1975, was the most profitable quarry for parrot smugglers. The last two offspring of the last pair of Lear's Macaws in the wild were stolen as nestlings from the hollow tree where they had been bred. The nest robber of the Rio São Francisco got ten thousand dollars from a middleman in the south of Brazil, who passed the precious creatures on to a wild animal dealer in Asunción, Paraguay, for twenty thousand dollars. The birds were to be shipped to West Germany via Switzerland. An importer there was willing to pay forty thousand dollars for the rare blue birds that would easily bring in eighty thousand from a private collector.

The documentation submitted for permission to import the Lear's Macaw chicks to Switzerland identified them as having been bred at the Asunción Zoo. An inquiry revealed this to be a sham, as the zoo had no Lear's Macaws at all. Uruguayan zoologist Juan Villalba-Macias, the watchdog responsible for trade in endangered species in South America, was alerted by Swiss authorities; he flew that very day from Montevideo to Asunción where, on March 25, he won over the minister in charge of implementing CITES[23] with a plan he had concocted.

At one o'clock that day, Villalba-Macias, with a search warrant in his pocket and accompanied by twenty police officers as well as the director of the Paraguayan National and Wildlife Parks, stormed the house of a wildlife dealer named Koopmann, whose daughter would be charged seven years later as Tony Silva's accomplice. The raid produced nothing, at first. No trace of Spix's Macaw nestlings. No cages, no bird feed in sight. Suddenly, the men searching heard a door slam on the second floor. Villalba-Macias ran upstairs and caught a maid who was about to fly the coop. In her travel bag were two downy Spix's chicks. "Tiny and

Christian Buhle, *Blue-and-yellow Macaw (Ara ararauna)* and
Spix's Macaw *(Cyanopsitta spixii)*, 1835; the Spix's Macaw was drawn according
to Spix's description, and the yellow color on its face is incorrect.

vulnerable, they were miniature versions of their parents—only with shorter tails, pale rather than dark bare faces and with a light stripe running the length of the hooked upper mandible. A few of their longer plumes were still partly sheathed in the hard plasticlike cases that would soon burst open to reveal their first full set of fresh blue flight and tail feathers."[24]

The birds were taken to the Brazilian embassy and flown to São Paulo the next day, where crowds of reporters, photographers, and cameramen were waiting for the little parrots and their rescuer. In a conference room, Villalba-Macias related the nestlings' family history to the media mob; all the while, the birds were strolling around on his desk. One of them pulled a rubber stamp off the carousel stand with its bill, the other, apparently bored by its rescuer's macaw family yarn, flew onto Villalba-Macias's shoulder and gently nipped his ear.[25]

Koopmann, a former military officer, wound up in jail, and Carlinhos, the bird trapper from Petrolina, the city across from Juàzeiro on the Rio São Francisco, was put on trial. However, the local judge felt it was his job to protect poor citizens from the "madness of environmental laws" and acquitted him.[25] Carlinhos confessed to having trapped eight Spix's Macaws since 1984. Where he did it, he preferred to keep to himself, in light of the fact that there were still three live birds there.

Since the two nestlings could not have found their way in the wild without their parents' guidance, they were taken to the São Paulo Zoo; three Spix's Macaws also confiscated from the black market were already there.

The last three Spix's Macaws—the parents of the two nestlings accompanied by an older male—were living in the vicinity of the area where Spix had come across his extremely rare macaws with the conspicuously thin voices. At the time he journeyed through Brazil, Juàzeiro was a little village of fifty houses and two hundred inhabitants. Today, over 128,000 people live there. But the vast country is as inhospitable as ever and therefore sparsely settled. Cultivated plants are no match for the infernal heat and aridity in the caatinga. During a seven- to eight-month drought, only shrubs and trees with bulbous, swollen root stocks or other water-storing cells are able to survive. These "impenetrable palisades," which Spix and Martius had ridden past, include the barrel tree with its barrel-like trunk and the "cactus tree, armed with long white bristles or menacing thorns."[26] "All of these plants are caatinga forms because they drop their leaves during the dry season and usually leaf out only after the rainy season begins. Only in wet lowlands do they keep their leaves year round; in other zones, the leaves so depend on moisture that at times two to three years must pass before the seemingly dead trees leaf out again."[27]

Gallery forests are the lone oases in the caatinga. They form parklike strips of trees and bushes along the banks of *riachos*. These "little rivers" only carry water during the rainy season, but plants draw the moisture they need from the deeper layers of fine sand in the dry stream-bed and the flooded bank areas, while animals slake their thirst with the brown water not yet evaporated from the dried-out *riachos*.

Martius called the large trees in the gallery forests bordering the little rivers to the northeast of Juàzeiro *Tabebuia caraiba* (Martius).[28] The branches of old Caraiba trees that had died after being starved by long droughts were like thin white fingers sticking out of the green treetops; Spix's Macaws used them as perches from which to look out for enemy raptors; hollows in the rough trunk were roosting and breeding sites, and the relatively soft-shelled Caraiba seeds, along with other seeds and fruits, were accepted as food. Rare the macaw, rare its habitat. The fertile soils of the river valleys were the only ones suitable for the irrigated planting of maize, soy, and sugar cane in the whole 310,000-square-mile wasteland. Once upon a time, the gallery forest on both sides of the middle section of the Rio São Francisco's tributaries ran thirty miles back into the caatinga. There are just over eleven square miles left, in three fragments.[29]

One hot morning at the end of April 1987, after a lavish breakfast of fruit, the last three Spix's Macaws went back to more distant fields in the gallery forest on the Riacho Melància and settled down in the crown of a Caraiba tree. After some careful preening and cheerful chatter in the shade of the waxy-leaved canopy, they were taking a nap when the shrill screams of a parrot in danger awoke them with a start. True to their parrot nature, they tracked the panicky calls to a bend in the stream and took up their post in a tree overlooking the dry streambed in which a distant relative of the parrot family was desperately beating its wings and squirming as it tried unsuccessfully to take off.

While the two young Spix's Macaws cautiously kept their distance, the older macaw did not hesitate to go down to the lowest branch nearest to the calling parrot. When two men from the bushes on the bank leapt out and pounced on it, it was just as unable to fly as the trapped decoy in the streambed.

The bird trapper and his helper cut off the branch the precious blue booty was clinging to. In a flash, the macaw and the branch were caught in a nylon net and stuffed into a wire-netting box in a Jeep whose owner was certainly poor no more. Passing the traumatized bird from smuggler to middleman to its final stop in a clandestine private aviary was carried off without a hitch.

Months went by. The first monsoonlike November rains initiated a new breeding season for the now extremely jittery pair of Spix's Macaws, still in their unchanging hollow in an ancient Caraiba tree on the Riacho Melància. From the middle of December, the female sat on three white eggs. On Christmas Eve, the birds were sleeping close to one another in their hollow, as always. Suddenly a noise jolted them from their slumber, the same noise that, in their superb parrot memory, was permanently linked with that disastrous April morning when their congener had been carried off in a roaring vehicle. The roar came closer and abruptly stopped. The ensuing rustling and scraping on the tree trunk, the whispering of the creatures they recognized as their worst enemies, filled them with terror. But it was risky to take off at night into owl territory. Better to pretend you didn't exist. Not until something pushed into the hollow did the female fly out, only to be entangled at once in a net. The moment she was caught, the entrance to the hollow was free for the male to make his escape.

The poachers blindly felt around in the hollow of the tree, expecting to find nestlings, but instead they came upon three eggs, which they broke with their greedy, fumbling hands. Dawn was breaking when they stuffed the box with the *Ararinha-azul* (Little Blue Macaw, as Spix's is called in Brazil) into the Jeep and covered it with an old blanket. They knew that Paul Roth, the Swiss ornithologist at the University of Maranhão, paid people around there to keep watch over the rare blue creatures. Because the thieves were working with armed accomplices, the people guarding the birds chose not to be visible. A chance passerby seemed harmless to the gang, so when he asked what was in the box, they showed him—and they were proud to do so. They even had no objection when he took a Polaroid picture of the bird behind the wire netting. Christmas was the ideal time for grabbing the bird. The guards had of course phoned Paul Roth the same day about the loss of the second-last Spix's, but the officials of the nature preservation bureau didn't hear the news until they came back from Christmas holidays in January 1988. By then the beautiful blue lady was long gone.

Roth heard that same January that the last macaw of the trio he had discovered on the Riacho Melància had been trapped. The population seemed to have been eradicated, but he could not believe that all Spix's Macaws had been wiped out. There could still be *Cyanopsitta spixii* somewhere, in some gallery forest of that giant country, he told the International Council for Bird Preservation (ICBP),[30] which had underwritten his search for the macaw.

A five-man team, including Cambridge parrot expert Tony Juniper, was organized by the ICPB to save the Spix's Macaw. In June 1990, the team reached the Minas Gerais region, where they heard the good news that Spix's had twice appeared in the Chapada das Mangabeiras mountain range the previous year. Although the team consisted of three ornithologists, a conservationist, and a wildlife photographer, it was only able to turn up Red-bellied Macaws in the scattered palm woods in the open grassland, birds that had apparently been confused with Spix's Macaw.[31] During their fruitless search in Minas Gerais, the team came across a bird catcher carrying Amazon parrot booty in three small cages on his bicycle.[32] "Spicara," he immediately named the bird in the photograph he was shown. His vehicle spoke for the credibility of his statement that there were no "Spicaras" in the Minas Gerais. You can only find them, he said, on the Rio São Francisco, almost four hundred miles east.

As the team made its way to the river, none of the farmers and hunters quizzed had ever seen a small blue macaw with a strikingly long tail. Juniper, depressed by the lack of success in his search, arrived with his Brazilian team in Curaçá, a small town in the caatinga on the Rio São Francisco, fifty-five miles downstream from Juàzeiro and only nineteen miles from the Riacho Melància. There they met a man who, the moment the word Ararinha-azul was dropped, produced a Polaroid picture showing the blue outline of a terrified Spix's in the shadow of the back wall of a box that had a wire netting in front.

The man had taken the photograph years ago at Concordia Farm near the Riacho Melància. The tenant was a young vaqueiro (cowboy) and his family. Before long, Juniper & Co. were hanging their hammocks in a vacant farm building without water or electricity, as paying guests of the wretchedly poor tenant and his pregnant wife. When asked about the rarest of all wild birds, she declared that Ararinha-azul could be seen almost daily in the tall trees at the bend in the river, where a puddle in the streambed never completely dried out.

The five men, armed with binoculars, a telescope, cameras, and video cameras, were lying in wait in the riparian brush overlooking this watering hole at four in the morning on July 8. No bird song could have made them happier than the *kraaa kraaa kraaa* of an approaching parrot at first light, greeting its green world on the Riacho Melància. Its trill might have sounded thin from a distance, especially when compared with the hard, harsh, or shrill shriek of other macaws, but it got fuller and louder as the bird came closer.

The last Spix's Macaw in the wild had not flown into a trapper's net. It flew in with deep wing beats among a flock of Blue-winged (or Illiger's) Macaws (*Ara maracana*, or *Propyrrhura maracana*) to drink at the river bend. Its blue silhouette, bright head, and sizable tail were conspicuous in the clear light of day compared with the smaller and shorter-tailed macaws' predominantly green feathers. Before the blue bird risked descending to the ground to drink, where snakes, wildcats, and other enemies might be lying in ambush for him, he stayed high up on the bare limb of a magnificent Caraiba tree, checking to see if the coast was clear. The cluster of his admirers under the foliage escaped his alert yellow eyes.

The last of the Spix's Macaws was continuously observed from that point on. The ornithologists were amazed to find that he had chosen a Blue-winged Macaw as his mate. He always flew next to his little green bird in their flock, unless he was alone with her when on the move or trying to mate with her. Whenever the latter happened, the female would flutter her blue wings anxiously. She seemed jittery. Something wasn't quite right about this larger bird, but she put up with this "nonspecific" lover nonetheless. She would stay with her flock only during the November breeding season, while her abandoned blue bird, in spite of his solitude, would defend his traditional breeding hollow in the Caraiba tree with furious screeching against pairs of Blue-winged Macaws seeking shelter.

The only way to enable *Cyanopsitta spixii* to survive in the Riacho Melància gallery forest was to reintroduce a captive female Spix's that had some experience in the wild. It so happened that the man who had sold the blue bird in the Polaroid photo was found. But years would pass

before the last bird in the wild would see his mate again. It's a tough battle to preserve such precious and famous creatures as the Spix's Macaw. If
the intention to go down in history as a Noah is more powerful than the
wish to save whatever can be saved, if people resent having other Noahs
around them, and if one breeder begrudges another his brood while an
unscrupulous macaw specialist secretly feathers his own nest, then the
ark is in a bad way.

It would have made the most sense to put into one aviary all fifteen
Spix's Macaws that were in human hands in various countries when
the last bird in the wild was discovered; this would have given the birds
a chance to select a mate. Genetic diversity would have been preserved,
and we would have been spared the laborious noninvasive process of
determining race. Not to speak of the much more complicated "mating"
by rival collectors. This would have presupposed that all Spix's owners
would understand that their birds, almost all of which had been taken
from the wild, were from caatinga gallery forests. But there was only one
person in the whole wide world who formally announced that his two
birds were Brazilian property: Wolfgang Kiessling, the founder and pro-
prietor of the parrot park on Tenerife.

The first meeting of the Comité Permanente para a Recuperação da
Ararinha-Azul was held in Brasilia in the middle of July 1990.[33] The
committee was made up of government officials, Spix's owners, scientists, and conservationists; it decided to work out a recovery program that
would optimize breeding with the aim of reintroducing the macaw into
the wild. A breeding master list was to be kept to prevent inbreeding.
Exchanges of birds scattered throughout the world, which was essential
for forming breeding pairs, was not to involve any financial transactions
by breeders, and trading in young birds was supposed to be banned. All
Spix's owners were to be contractually bound in this manner. It remained
unclear whose birds were to be reintroduced and to whom the young
birds belonged if a pair shared by two owners incubated only one or
three chicks.

Brazil was given authority to confiscate all illegally held Spix's Macaws. Tedious lawsuits were incompatible with the recovery program, and so the government declared a five-year amnesty in February 1991 for those who signed the contract agreeing to the permanent committee's conditions. Subsequently, more birds surfaced in Switzerland. Dr. Hämmerli had purchased his first Spix's Macaws for his personal aviary in 1978 and acquired, via Koopmann and his fake export documents, his two most recent macaw chicks: the second-last brood of the last pair of Spix's Macaws on the Riacho Melància.

The committee was split on what to do with the last bird in the wild. Some were in favor of catching it to save its genes, arguing that maybe even tomorrow it could fall prey to a raptor, a wildcat, a snake, or a swarm of Africanized honey bees.[34] Tony Silva was with this group. He had not yet been unmasked and knew what he was talking about when he argued that the bird might wind up in a trapper's net. Others countered that reintroducing aviary birds, most often hand-reared nestlings, without a wild bird to guide them was doomed to failure. All hopes for saving the bird were to be pinned on a single bird that still knew what it meant to be a Spix's Macaw in the wild. This faction carried the day at the end of 1992.

It was high time to get to work on determining the sex of the macaws in captivity. Appearance alone was not to be trusted, even in the case of a wild bird. There must not be a repeat of what happened to a breeder in 1978, who borrowed a male Spix's to breed with his female. He was filled with cheerful expectation that there would soon be eggs after seeing how the loving pair dug a proper breeding hole in the trunk of a palm and, as per the parrot way, copulated. When no eggs appeared, an endoscopic examination found that both birds were male.[35]

Meanwhile, molecular biologists had learned to sex parrots by analyzing genetic material taken from shafts of feathers plucked less than twenty-four hours earlier. They only had fallen-out feathers from wild birds, which could only yield DNA fragments. It took eight months to develop a new technique for reproducing the fragments. The procedure then had to be tested with material from known species related to Spix's Macaw, which took many more months. Finally, the first experiment using minimal material from three of the wild bird feathers revealed that it was a female. It was some time before the field scientists at the Riacho Melància collected a fairly large number of molt feathers. Almost two years passed before it was finally determined, in January 1995, that the lone Blue Macaw in the gallery forest actually was a male.

He had been alone for seven years. It was time to provide him with a female. His beautiful lady Blue-winged Macaw had wound up in the bird breeding center of a trucking operator in Recife. It was in the company of a Spix's Macaw male from the São Paulo Zoo, presumably a son of hers. On the initiative of the permanent committee, they were both taken in August 1994 to an open-air aviary near Concordia Farm that had been built with donations from the Loro Parque Foundation.

The female Spix's Macaw's behavior in the sixty-five-foot-long enclosure, with a stimulating view of the green oasis she had come from, reinforced the scientists' hope for a successful reintroduction. In the compound's trees the female was able to show the bird, who had been moved to Asunción as a nestling, how to dive for cover if a large raptor were circling overhead, how to peck wild fruits and seeds with one foot and open them expertly with its bill, and how to answer the trilling *kraaa, kraaa, kraaa* from outside the oasis. The supposed son never mated with his mother during the four years they were together, but the birds were very attached to each other.

Just the female was to be reintroduced. As long as she had not found her bearings in the wild, she would look for the bird in the aviary and not go astray. But what would become of the liaison of the Spix's with the Blue-winged Macaw? Parrot pairing is permanent, as a rule, but mating between Spix's and Blue-winged Macaws normally does not occur. After endless scientific debate over whether the blue-winged green bird that was blocking a marriage between the Spix's Macaws should be caught or even shot, it was agreed to let nature take its course. The risk of traumatizing the wild bird was too great.

Before the tame Spix's was released to its feral male, its tail was shortened a little so that an observer could distinguish the two. The aviary door was opened for her on the morning of March 17, 1995. A short flight, and she was perched again in a tall Caraiba tree near the riverbed, staying a while in the shelter of the green canopy and exchanging calls with her species mate still in the aviary, and then flew back to him.

Soon she could be seen flying with deep wing beats on longer excursions. It took a few weeks before she was completely at home in the wild and gave up the visits to the aviary. At first, her flight paths only crossed with the male's, but soon the two Spix's Macaws were flying together through the gallery forest, sometimes with and sometimes without the little Blue-winged Macaw. They could be seen billing and cooing in May. It didn't seem to bother them that the Blue-winged Macaw usually stayed nearby.

Everybody played a role in the fate of the last Spix's Macaw. The media compared them, reunited after seven years, to Adam and Eve because all future generations would proceed from these last individuals just as they had done from the first. But the caatinga people saw *Ararinha-azul* as a "local hero." The scientists monitoring the last surviving Spix's Macaw had converted the people to their preservation program. The people of this godforsaken region could easily identify with the lone bird that callous profit-seekers had not trapped and that had eked out its survival in a bleak natural setting.

The recovery program began during the most terrible drought in the caatinga in living memory. It was evident that even the most beautiful bird of all was beyond help as long as the people around it were starving to death. Seven tons of food was donated in the name of the permanent committee. Concern about the macaw thus became concern over the people around it.

Stickers and posters depicting the Spix's Macaw were distributed on market days so that even the most remote farmer knew the number to call in Curaçá or Brasilia to pass on information. Over time, Curaçá and its surroundings were seized by outright Spix's Macaw fever. Curaçá's name became Ararinha-azul City, where there was, naturally, a Spix's Macaw Restaurant. Actors in blue bird costumes in an old, run-down theater performed a Spix's Macaw play recounting the bird's longtime loneliness, its desperate liaison with the little Blue-winged Macaw, its mate's imprisonment, its return, and the happy ending in the gallery forest with a Spix's flock. Wolfgang Kiessling and other committee members saw a performance, after which they championed the restoration of the small theater, which was made possible by a donation from the Loro Parque Foundation. And an *Escola da Ararinha* for twenty-two children in the area was built on the Riacho Melància with the committee's support. The pupils helped put up fences to keep sheep and goats from grazing on Caraiba saplings. Saving the Spix's Macaw required saving its habitat.

More than a hundred caatinga *vaqueiros* helped guard the two blue birds and the green third bird. Their observations supplemented those of the scientists. A change of heart on the male's part occurred in June 1995. The assiduous loyalty of the little Blue-winged Macaw had its effect on the beautiful female Spix's, who was not about to look on as her darling mate paid more and more attention to the Blue-winged female. She flew off and was not seen again. A "Spix's Macaw *vaqueiro*" found her dead under a high-voltage line but feared this information would endanger the recovery project. He only disclosed what he knew four years later.

Now the Spix's was once again to sleep on a tall, thorny cactus while his bride spent the night with her own kind. The odd couple could be seen in the daytime flying together to their food trees. As the breeding season began, the lone bird had to defend the entrance to his Caraiba nesting hole once more against all sorts of birds. In the meantime, the female stayed with the Blue-wingeds. It would take a year for her to move into the Spix's hollow. Researchers found three eggs there in December 1996, which they switched with Blue-winged Macaw's eggs from another nest because they were certain that the three eggs were infertile. As expected, there were no hybrids among the hatchlings in the incubating box. But an investigation showed that a ten-day-old embryo in one of the eggs had died, and its DNA proved that it was from *Cyanopsitta spixii*.

The female Blue-winged Macaw was tended to by the Spix's and brooded the little Blue-wingeds. Should the pair succeed in raising its brood, Spix's Macaw nestlings would be placed under the female during the next breeding season. But an opossum or a common marmoset gobbled up the delicious Blue-winged chicks, and the scientists had to wait a full year to repeat the experiment. In December 1997, a snake probably found the Blue-winged eggs in the Spix's nesting hole simply delectable. The next season, the nesting site was shielded from apes and opossums with a metal collar around the tree trunk. To prevent any loss, scientists no longer replaced the pair's sterile eggs with fertilized ones but with wooden imitation Blue-winged eggs instead.

After incubating for twenty-three days in January 1999, the Spix's Macaw's green mate returned to the nest for a quick snack; in the meantime, two peeping pipsqueak Blue-winged Macaws had hatched from the silent eggs, leaving no trace of a shell behind. The two little ones would want for nothing from then on. Their supposed progenitors were exemplary parents. The fledgling Blue-wingeds followed their blue father and their green mother into the gallery forest in March. Their Spix's calls came across to the scientists as a good omen for raising future Spix's changelings.

The permanent committee planned a trial reintroduction on the Riacho Melància with a less endangered macaw species in order to reduce the risk of releasing tame Spix's Macaws to a minimum. To this end, in early 1997, the Loro Parque Foundation donated twenty young Blue-winged Macaws raised on Tenerife and had them shipped to Recife. The

Anita Albus, *Blue-winged Macaw (Ara maracana)*,
Spix's Macaw Female and Male on a
Caraiba Branch, watercolor and body color on paper, n.d.

birds' acclimatizing quarantine at Mauricio dos Santos's breeding center was a disaster.[36] They were kept in much too small wire cages and began mutilating themselves. After the promised large aviaries were still not finished months later, Kiessling sent his veterinarian to Recife. Eight of the nineteen Blue-winged Macaws he found were in such a pitiful state that they could only be saved on Tenerife. The other eleven were able to recover from their wounds from November on in the open-air aviary on the Riacho Melància.

The eleven birds in the large compound had scarcely recuperated when they were demoralized once again. They were decked out with newfangled transmitter necklaces that were supposed to guarantee that they would be monitored in the wild for two years. Parrots had not yet been tested to see if they would accept electronic instruments around their necks. The results of the experiments in the compound were devastating: "The maracanas became pathological. Some of the pairs broke up, dominances in the group changed, two of the birds were killed in fights, one by its own partner, and a couple of homosexual pairs formed."[37] The birds ultimately destroyed the expensive transmitters.

Loro Parque conducted a test in September with lightweight transmitters on the macaws' tails. The birds tolerated the light baggage, and the scientists felt that the transmitters would last six months. The nine macaws left were equipped with the transmitters. Only one partner of each pair was released at first so that the birds would stay around the compound, where water, feeders, and nesting boxes were provided. After this successful cautious reintroduction, the last four Blue-winged Macaws were released into the gallery forest in January 1998.

Not all nine birds were expected to survive in the wild, but seven did make it. Their success spurred the Spix's rescuers on. The breeders also had good news. There were now over sixty Spix's Macaws in their aviaries. A breeder in the Philippines was prepared to volunteer two males and three females for reintroduction. It took a while to get the five macaws to the Riacho Melància compound to join the supposed son of the Spix's female, which had disappeared into the wild. Only smugglers were able to get export papers for rare birds virtually overnight. The permanent committee made plans in the meantime for further swappings of change-lings in the Spix's nest hole. During the 1999–2000 nesting season, Blue-winged chicks were once again foisted on the blue-green pair, since there were no little Spix's Macaws there. The chicks flew out successfully and were taught their Spix's lessons by their father. The next breeding season was to see the switching of tried-and-true wooden eggs with Spix's nestlings after a twenty-three-day waiting period.

If the food trees on the Riacho Melància were exhausted after a long drought, it was possible for the numerous wild Spix's observers to lose sight of the bird for a week or two. It was still seen on October 5, 2000. Two weeks had gone by since then. Days were counted, soon weeks. Search parties combed every corner; every farmer was questioned, but no one had seen *Ararinha-azul*. Meanwhile, the green mate of the van-ished bird was still waiting in the neighborhood of the breeding hollow. It seemed not to want to believe that the big blue bird had disappeared once and for all, any more than people wanted to. The Spix's had fended for itself in the wild for thirteen years, alone. It was never seen again, either dead or alive.

The Spix's Macaws selected for release stayed in the Philippines. But the breeders there had, behind the committee's back, sold off four birds to a sheikh in Qatar. Hämmerli didn't keep his word either and had "hawked" his Spix's Macaws in Switzerland; authorities there felt no compunction to intervene. The ornithologist at the Houston Zoo who was in charge of keeping breeding records supported the deal with the sheikh, who was lobbying to be put on the committee only if there were no strings attached. There was one more meeting in Brasilia in February 2001. The Loro Parque Foundation had given $600,000 to the recov-ery project. Reason enough for some envious members to leave the room when Kiessling came in.

Once *Cyanopsitta spixii* had disappeared in the wild, the committee to save it disbanded. Three years have passed, and sixty Spix's Macaws are still waiting in various aviaries to fly like their ancestors in flocks through the gallery forests. The shrewd Noahs have not vanished from this earth; the oasis on the Riacho Melància is still there, as are the Caraiba trees orphaned by Spix's Macaw, trees that unfold their yellow trumpet flowers from August to September. Will the hollows in their chapped trunks ever again harbor blue nestlings? The answer is written in the stars.

PART II

Birds Threatened
and Endangered

Conrad Gesner, *Waldrapp*
(Geronticus eremita), 1617

{ CHAPTER 5 }

The Wondrous Waldrapp

AFTER GOD created the world, He brought all the animals of the field
and all the birds of the air "unto Adam to see what He would call
them: and whatsoever Adam called every living creature, that
was the name thereof."[1] Nibbling on fruit from the Tree of Knowledge
put an end to the beautiful accord between divine creation and human
naming. Since Adam and Eve's expulsion from Paradise, naming animals
has had its pitfalls.

No forest has ever seen a Waldrapp; no bird in the raven family has
ever recognized *Corvus sylvaticus* as being of its ilk.[2] Even in antiquity,
the Crested Ibis (*Schopfibis*) was insulted by being called *Phalacrocorax*,
"the bald raven," a name later transferred to the cormorant, whose "raven-
ness" was also restricted to its black feathers. *Erdhuon* was the ungainly
Old High German form of the Greek and Latin word *ibis* found in
twelfth-century manuscripts. Albertus Magnus mentions a *Corvus terre-
nus* a century later in the book of his *Liber de animalibus* devoted to birds,
and who could this "terrestrial" raven (*Erdrabe*) be but our Waldrapp?[3]

The Waldrapp's most felicitous designation also stems from the thir-
teenth century and is found in a work by a heretic whom Dante consigns
to the flames in the sixth Ring of Hell: *De arte venandi cum avibus*, "On

the Art of Hunting with Birds," by Emperor Frederick II. Observing the birds' various flight patterns, he compares the Waldrapp, named *Galeranus campester*, with the Glossy Ibis, *Galeranus aquaticus* (now *Plegadis falcinellus*). In the chapter "On the Use of the Parts of the Body and Their Differences in Individual Avian Species," he compares *Galeranus niger campester*, which he states has no feathers or down on its head, with the bald head and neck of the Sacred Ibis, which he calls *Galeranus varius ex albo et nigro*. The ibislike nature of the Waldrapp had not escaped this ornithologist's sharp eye. His fortuitous association of the Waldrapp's feathery decoration with the classical helmet (*galea*), evident in the term *Galeranus*, did not resurface in species terminology until centuries later as *cristata*, "wearing a panache." The Glossy Ibis has closely flattened head feathers resembling a metal-studded visor on a helmet because the white-fringed dark brown feathers reflect light and because of the cut of its narrow white mask running from the eyes down to the bill. The sacred *Galeranus* has to get along without any decoration on its velvety black helmet, which goes halfway down its neck. The Hohenstaufen emperor knew from his crusade to "the Orient" (i.e., the Near East) that the Sacred Ibis "is frequently found in Syria, Egypt, and the Far East."[4] The Glossy Ibis had a home in the emperor's residence. The chestnut-brown *Galerani aquatici*, whose wings had a metallic sheen in the sunshine, would have visited the marsh garden and its ponds, which the Hohenstaufen emperor laid out for his study and enjoyment. Symbolic of Frederick was "his large vivarium ... Close to Foggia he had a big marsh, laid out with ponds and walled water conduits, which was alive with all descriptions of waterfowl. A fantastic picture—the great palace with its columns of marble and serpentine, with bronze and marble statues, the Emperor within attended by Moorish slaves and noble pages, visiting his pools to study pelicans, cranes, herons, wild geese and exotic marsh fowl!"[5]

As its name says, *Galeranus campester* does not belong to this aqueous avian society. It does indeed search out rivers for drinking and bathing

Der braune Ibis.

Christian Buhle, *Glossy Ibis*
(*Plegadis falcinellus*), 1835

and does not reject small fish and crustaceans either; but its main prey—snails, lizards, beetles, scorpions, grasshoppers, etc.—are found in wild areas with sparse vegetation. The "divinely sublime Emperor of the Romans, King of Jerusalem and Sicily," must have observed the bird's "rapid" flight over a dry steppe with "rather few wing beats" alternating regularly with periods of gliding. He apparently did not know the Waldrapp firsthand, or else he, who was concerned with "depict[ing] things as they are," would not have described the bill and legs of *Galeranus niger campester* as black. Nor does the bird appear in his falcon book as game.

Three hundred and forty-six years after its author's death, a print version of *De arte venandi* appeared in 1596 in Augsburg, edited by Markus Welser from the Manfred manuscript (annotated by Frederick's son Manfred) belonging to Joachim Camerarius the Younger. Included were Frederick's findings—always far ahead of their time—on birds' lives, their behavior, breeding, flight, types of migrations and their causes, molts, senses, anatomy, and the form and function of their diverse feathers. Of all the great sixteenth-century natural scientists, only Ulisse Aldrovandi was able to make use of the falcon book for his three-volume *Ornithologia*. Camerarius, a former student of Aldrovandi's in Bologna, had sent him a copy of the first edition. The book's publication came too late for William Turner, Belon, and Gesner. They, too, were working on exact descriptions from nature. Had they still been living, they would not have failed to read their dauntless predecessor's book.

William Turner's reference to the Waldrapp in his *Avium Historia* is so skimpy compared with his other meticulous descriptions (for instance, of the Red-backed Shrike), it's as if he were basing his description on hearsay. In his book, published in Cologne in 1544, Turner, a humanist and a friend of Conrad Gesner, calls the Waldrapp *Helvetiorum Vualtrapus*. Pierre Belon's *L'Histoire de la nature des oyseaux*, published in Paris in 1555, portrays a Glossy Ibis under the name *Ibis noir*. But Belon's description of this "Black Ibis" nearly matches the Waldrapp but for the missing

Pierre Belon, *Glossy Ibis*, 1555

feathered crest: "This ibis of which we wish to speak is barely smaller than a curlew and totally black; its head is like a cormorant's, its thumb-thick bill is pointed, curved with an upward hump, and completely red as its thighs and legs; its height, look, and attitude resemble a bittern's."[6] On Belon's long journey to the Near East, he appears not to have seen a Waldrapp—he wouldn't have missed that shock of feathers. His description points more to a somewhat forgotten Eastern specimen. At any rate, the ibis shows neither the cormorant's bare spots on its face and neck nor a red bill and legs.

No ibis, black or white, figures in the woodcuts in Conrad Gesner's famous bird book of 1555, the third volume of his *Historia animalium* that Froschauer published in Zurich. Yet, Gesner's description of the birds

Pierre Belon, *Phalacrocorax*, 1555

from "the land of Egypt" is rich in detail. The Swiss scholar writes, "The black [bird] has long legs, and is as large as the bird called *Crex*," and he puzzles over the bird's possible relatives in his own country: "In the Alps one finds a bird, named a black stork, that cannot be a black *Ibis*, because of its straight bill nor can the 'Waldrabe' be the black *Ibis*, although it has a curved bill, whereas it does not resemble it in any other way." He almost recognized the wrong raven. And he hits the nail on the head when he goes on to talk about *Plegadis falcinellus*, the Glossy Ibis, as it's called today: "That bird can well be somewhat related to the *Ibis*, which the Italians call *Falcinello*, which we will describe to be like a heron; it has a long and bent bill like a bow or a sickle, and its feathers are of a green color."[7]

Belon's and Gesner's reports are based on Herodotus, Aristotle, and Pliny, who said that the black ibis only stayed near Pelusium in the eastern part of the Nile Delta, whereas the white Sacred Ibis was found in all of Egypt. Gesner gives a vivid retelling of the ancient legend of the ibis as the discoverer of the enema: "This bird, when its body is constipated, gives itself an *enema* so: It fills its throat with sea water behind and

DE FALCINELLO.

Conrad Gesner, *Glossy Ibis*, with head of a heron, upper left, 1617

into itself, and purges itself and eliminates thus by the pungency of the salt water. Which is why some claim, that this bird and the heron first brought about and invented the *enema*."[8]

In the Alps, this feat was not attributed to the Waldrapp, and so Gesner's description of the local ibis had to manage without any reference to legend. Apart from the matter of the woods and ravens, there was no lack of precision in his report:

The bird of which the figure is here given is generally called by our people a Wood-Raven ("Wald-Rab"), because it lives in the uninhabited woods, where it nests in high cliffs, or old ruined towers and castles, which caused it to also be called Stone-Raven ("Stein-rab"), or elsewhere in Bavaria and Styria "Klausrab," ["mountain retreat-raven"], from the rocks and narrow caves and holes in which it builds its nest. In Lorraine and around on the Lago Maggiore it is called a Sea-Raven ("Meer-Rab"); in other places Wood-Raven, as in Italy, where a man is lowered down on a rope to take it out of its nest, as it is considered a great delicacy ("ein schleck"). In our country it is found also in the high cliffs near Pfaffers, where some hunters went down for it on ropes.

From its voice it is also called a Ringer ("Scheller"). Some authors take it to be the *Phalacrocorax*, for in size and color it resembles the raven. It acquires also a bald head in its age, as I have seen. Turnerus takes Aristoteles' Water-Raven, Plinius' *Phalacrocorax*, and our Wood-Raven for the same bird, but it is wrong, because their descriptions are unlike, the Wood-Raven not having broad feet and not being a water-bird, but seeking its food in green meadows and swampy places. Our Wood-Raven is of the size of a hen, quite black if you look at it from a distance, but if you look at it close by, especially in the sun, you will consider it mixed with green. Its feet are also somewhat like a hen's, but longer and the toes split. The tail is not long. It has a crest on its head pointing

Conrad Gesner, *Cormorant (Phalacrocorax)*, 1617

backwards, though I do not know whether this is seen in all individuals and at all times or not. The bill is reddish, long, and suited to poke with it into the ground, and into the fissures and holes of walls, trees, and rocks, to extract the worms and beetles which hide themselves in such places. Their legs are long and of a dark red. They live on grasshoppers, crickets, small fishes, and frogs. They generally nest on the high old walls of the ruined castles, of which so many are found in Switzerland. When I dissected their stomachs I found, among other vermin, also many creatures which are injurious to the roots of agricultural plants, especially the millet. They also eat the grubs which produce the cockchafers. They fly very high, and lay two or three eggs. They fly away first of all birds, really in June, or as others told me about St. Jacob's Day. They fly in swarms and cry "Ka, ka," and most of all when their young are taken, which is generally done about five days after

Whitsuntide. They return to us in early spring, when the storks arrive. If the young are taken from the nest some days before they fly, they may be easily reared and tamed, so that they fly out to the fields and quickly return. The young ones are also praised as an article of food, and considered a great delicacy, for they have a lovely flesh and soft bones. But those who rob their young leave one in every nest, in order that they may like to return in the following year.[9]

Ulisse Aldrovandi, *Waldrapp*, 1599–1603

Nest looters and hunters condemned the Waldrapp to rarity status in the sixteenth century. The Little Ice Age took care of the rest because there were fewer and fewer places for its beak to probe. By the time Ulisse Aldrovandi's third volume of his *Ornithologia* appeared in Bologna in 1603, most of the Bald Ibis's rocky retreats in Europe might well have been abandoned. The model for Aldrovandi's 1603 likeness of the bird had been sent from Dalmatia. There could not be a more ibislike portrayal of the Waldrapp than in Aldrovandi's *Phalacrocorax*. With its bald head thrown back into the mane on its neck and its sickle bill craning up to heaven as if summoning the gods to witness his triumph, the victor over evil holds a twisted serpent in his claws. Aldrovandi missed the connection between the bald-headed Dalmatian bird and the bald bird revered in Egypt that he had painted. He mistook the *Phalacrocorax* for a species of cormorant.

Thus, the white ibis with the bald black head become famous not as the inventor of the enema but as the serpent exterminator and earthly embodiment of Toth, or Thoth, the moon god and symbol of Egypt; it was thought that the country would be lost without the sacred bird's protection. Its reptile-taming power was said to extend to the tips of its magnificent plumage and the blue-purple sheen on the tatty scapulars of its shoulders: with the touch of an ibis feather, a snake or even a crocodile would lose the power to move. Ibis meat was viewed as poisonous because of the bird's snake diet, and the basilisk supposedly hatched from an ibis egg. Whoever brought about an ibis's death, even if by accident, paid for it with his life.

Belon, too, passed on the ancient lore of the Sacred Ibis's legendary serpent diet. The bird, esteemed in Egypt for liberating the country from the plague of snakes, was said to gobble up snakes wherever it found them and to attack them even when it was sated.

At one time, the measured stride of an advancing Sacred Ibis was omnipresent in Egypt. Today, it lives there only in the moon. Westerners

Christian Buhle, *The Sacred Ibis*
(*Threskiornis aethiopicus*), 1835

see a man in the moon; people in India see a rabbit; Egyptians see an ibis. Thousands of the animals sacred to the ibis-headed Toth have survived for ages—as mummies, with their heads and necks tucked into the shape of a heart—in the tomb of the vizier Imhotep found in a pyramid at Saqqara.

THE LOST WALDRAPP

The Waldrapp had long been extinct in Europe when the copper engraver and watercolorist Eleazar Albin painted a "Wood-Crow from Switzerland" in 1738 for the third volume of his *Natural History of Birds*. Soon after the Waldrapp became extinct, stuffed specimens became rare as well. The specimen that Albin laid eyes on seems to have been one of the last. Having escaped the museum's beetles and moths, which consider all specimens a "delicacy" and could not be defended against in those days, the Waldrapp specimen, along with an entire collection of other specimens, fell victim to a fire a short time later.

In 1758, Carl von Linné (or Linnaeus) catalogued the feather-crested and probe-billed bird depicted in Gesner and Albin among the hoopoes. This wondrous hoopoe seemed to live such a hermit's life in the impenetrable retreats Gesner described that no mortal had ever managed to see it in more than one hundred years. Linnaeus named the species *Upupa eremita*, after the solitude of its nesting places. But even that self-styled eavesdropper in "God's secret council chamber"[10] was not inured to doubt. In 1766, Linnaeus bowed to the traditional designation of crow: *Corvus eremita* was to be the bird's name.

Mathurin-Jacques Brisson wasn't buying the revered Swede's *Upupa eremita*. He had already transformed Linnaeus's categorization of birds into six orders and eighty-five genera into a more elaborate system of 115 genera in twenty-six orders. Even Pierre Barrère's "Wood-curlew" (*Arquatus sylvaticus*)—Barrère's taxonomical bird charivari *Ornithologiae Specimen Novum*, which came out in Perpignan in 1745 and made a mockery of any attempt at ordering nature—didn't make any sense to Brisson. In Brisson's six-volume *Ornithologia*, printed in Paris in 1760, the Waldrapp, now placed among the crows, appeared as an Alpine crow with panache: *Coracia cristata*.

The question over whether Gesner's *Corvus sylvaticus* was a crow, a hoopoe, or a curlew could not be answered without a living specimen or a Waldrapp skin. Not even the discovery of the South African Southern Bald Ibis (*Geronticus calvus*), painted by Georg Forster in 1772 as "the bald-headed ibis from the foothills of Good Hope" on the Cape, was able to shed light on the puzzle. Not one person suspected that this was the only bird in the same genus as the Waldrapp. Ibises were either regarded as storks going by the name of *Tantalus* or assigned to the curlews. And

Georg Forster, *Southern Bald Ibis (Geronticus calvus)*, 1780

so Buffon baptized the bald bird from the Cape *Courlis à tête nue*, whereas Johann Reinhold Forster, Georg's father, christened it *Tantalus capensis*. Linnaeus had already assigned *Tantalus ibis* to the African Yellow-billed Stork, which was vaguely reminiscent of an ibis and belongs today to the family of storks, as *Ibis ibis*, while ibises are now called *Threskiornithidae*, "sacred birds."

If, as the Chinese proverb goes, the beginning of wisdom is to call things by their *right* names, then wisdom was squandered in the naming of several genera and species. As long as palpable evidence was a criterion for naming, ornithologists were continually squabbling. Depending on the school or taste, they saw this or that name as being more appropriate. Some called storks *Tantalus*, others *Pelicanus*, and still others *Ibis*. The confusion did not end until the law of priority was established. Names are empty words; it all depends on uniformity, not on semantic correctness. Whoever is the first to describe and christen a genus will not be listed by name among zoologists unless he also names a species. According to the law of priority, the honor of having one's name attached to the beast is reserved for the person who first describes it, even if his nomenclature is highly misleading.

The bird Buffon described in his *Natural History* as the Sacred Ibis was a Yellow-billed Stork. Georges Cuvier corrected the error after he examined mummified ibises from Egypt. This is how the ibis family came to be separated from the storks. Although either bird could, originally, have been called *Ibis* or *Tantalus*, the latter term was reserved for the storks after the family was split. Because the law of priority was strictly followed, the ibis error was, in the end, preserved in the name for the Yellow-billed Stork. If it was mistaken at first for an ibis, the genus still had to be called *Ibis*. So the ibises lost their name, and the Yellow-billed Stork was turned into *Ibis ibis* Linnaeus.

Decades after the bald Cape ibis was discovered, the Berlin naturalists Friedrich Wilhelm Hemprich and Christian Gottfried Ehrenberg were traveling along the Red Sea coasts. On the Saudi Arabian side, some hunters shot two black ibises for them that differed from the South African species only by their long neck mane. The two skins entered the Berlin Zoological Museum in 1825 as *Ibis comata* Ehrenberg. After Hemprich's death in Eritrea that same year, Ehrenberg named the bird *Ibis hemprichii* in 1832, after his friend, but his work remained unpublished, so the name was never validated. It wasn't until thirteen years later that Eduard Rüppell portrayed and publicized the Waldrapp as *Ibis comatus* Ehrenberg. In 1849, the genus name *Geronticus* was applied to it, and it became known as *Geronticus comatus* Rüppell; Johann Wagler had introduced the genus name in 1832 for the South African Southern Bald Ibis. Meanwhile, a steady stream of Waldrapps was discovered in Algeria, Morocco, Ethiopia, Syria, and ultimately in Bireçik, Turkey. Still, the fact that the Waldrapp and the "Wood-Crow" were identical didn't dawn on any ornithologist. The latter had been considered all this time to be Gesner's invention, and his work was hardly considered required reading anymore. Nevertheless, ornithologists should have known that there was no way Gesner could have had the skin of a half-moldy Alpine crow with disfiguring molting head feathers before him when writing about the Waldrapp.

Ready to disbelieve the Waldrapp's existence, the ornithologists shrank from striking the bird from their classification scheme. There were, though, a few loyal "Waldrappers" who thought the feathered hermit might be nocturnal and could have been leading a life hidden from man since the sixteenth century. They, too, hadn't read their Gesner.

Finally, one day the scales fell from an ornithologist's eyes when he aligned an image of a Waldrapp with that of a Crested Ibis. The illustration that triggered this alignment was Eleazar Albin's Waldrapp painting that Henry Eeles Dresser included in his *History of the Birds of Europe*, which appeared in fascicles between 1871 and 1881. The eyes that at last recognized the new ibis in the old crow belonged to three ornithologists:

Johann Matthæus Bechstein, *Waldrapp*, 1809

Lord Lionel Walter Rothschild, the owner of the zoological museum in Tring near London; Ernst Hartert, the museum's director, and Pastor Otto Kleinschmidt, a naturalist from Wittenberg. The trio based the identity of the two birds on the evidence of a skin from Bireçik and published their identification in Rothschild's journal, *Novitates Zoologicae*, in 1897.

THE WALDRAPP BOWS IN GREETING

The Crested Ibis was protected as a semisacred bird in many Near and Middle Eastern countries. To Muslim eyes in Bireçik, the black ibises with their iridescent plumage carried the souls of the deceased. Since the birds went south in great numbers for the winter, just like pilgrims to Mecca, they were regarded as leaders of the Hajj. When the birds returned to Bireçik in the middle of February, they were revered as the harbingers of springtime. The day they arrived was celebrated as the anniversary of Noah's rescue from the Flood. The Turkish version of the Flood had a Crested Ibis showing Noah the way from Mount Ararat down into the fertile Euphrates valley, where he was to build a "little house," a *Bir-evçik*. To mark the "baldies'" arrival—they are called *kelaynaks* in Turkish—every year the Euphrates ferrymen would throw a lavish feast, where the poor were treated to the meat of the choicest lambs.

In Syria, where there were large Crested Ibis colonies, the birds did not receive a festive welcome upon their return up the Euphrates from their Ethiopian winter quarters; but, hungry Bedouins aside, they had little to fear as semisacred birds. The nomads indeed regarded them as they did other fowl—as a "delicacy"—and hunted them; the birds were safe from nest robbers only if they nested on the steepest limestone cliffs. These losses would probably have been compensated for by their strong population numbers if they had not been identified as Waldrapps. Once word got around, they were persecuted without mercy. As Crested Ibises, they had been, in the worst case, a delicacy; as Waldrapps they were a treasure, the best the avifauna in the countries on the Euphrates had to offer. The local hunters and egg thieves were joined by ornithologists and other "bird lovers" who carried off eggs, fledglings, and skins of the legendary ibis—a bird that had once bred in Austria and Switzerland, in Germany and Dalmatia—for zoological collections and zoos.

Plundering Waldrapp nests was now a lucrative business even for Bedouins. The zoologist Israel Aharoni was outraged by the "haggard, emaciated Bedouin thugs," comparing their hordes around the Waldrapp nests to swarms of blow flies. Aharoni filched over a hundred eggs and thirty fledglings from the last Syrian ibis colony near Palmyra, sending them to zoos in Europe along with nearly a hundred skins of Waldrapps that had been shot.

In Syria, too, the Crested Ibis once lived in close proximity to humans. European travelers in 1836 could still view thousands nesting in the city walls of Rakka on the Euphrates, the Nikephorian of antiquity. Persecution had taught the Waldrapp to flee humans. The treeless and shrubless plains of their feeding grounds offered hunters no chance of approaching the birds undetected. But not only the Waldrapp's large flight zone made it difficult to bring them down, either in flight or while foraging. Even if a shooter successfully hit a bird from long range, the bullet would bounce off feathers that were hard as a suit of armor.[11] Only breeding birds would not fly off until the enemy was fairly close by. After Aharoni had showed no qualms about popping off breeding birds at the nest and panicking the whole colony, he lamented their demise in 1929. By then, every colony in Syria was extinct.

Meanwhile, the black-crested soul-bearers of the Bireçik dead were fêted, as before, as Noah's messengers and the harbingers of spring. Soon after the popular welcoming festival to celebrate the birds' return, the ibises' bowing ceremony would begin. The birds would circle the breeding cliffs for two or three days; then they greeted each other with loud *chrups* before mating on rocky ledges and in niches on cliffs. If Linnaeus had heard these birds, he might have named them *Chrup chrup!* With heads thrown back and manes erect, males and females begin greeting one another in pairs. They hold that pose for a moment. Then, while giving its call, one bows deeply and vehemently toward its opposite number, showing off its headdress, which identifies each bird individually. A brief interval suffices for the bird on the receiving end of the bow to read the signs. Then the bowing bird throws back its shaggy head and repeats the welcoming bow several times in succession. The bird's opposite number will respond with a similar greeting. One pair's welcoming duet initiates a chorus of bowing Waldrapps on all sides, conducted by an invisible hand. If a female is unwilling to mate and flees the bowing male, the scorned bird chases after the escapee, bowing repeatedly and *chrupping* on the double.

The Waldrapp's bowing rituals are not limited to mating season. All the important events of the bird's life begin and end with a devoted greeting bow. Males greet females the same way they greet one another, though females are less welcoming among themselves. Bowing is foreplay to the invitation to copulate. Bowing precedes the attitude the birds assume to announce to each other that they want to be stroked with a bill; with a bow, the male points out the nesting site to the female; amid bows, they present each other with nesting material, and amid bows, they take turns at the nest.

The invitation to copulate is characterized by a symmetrical bill-in-bill figuration. Bills tightly locked in an *x*, male and female stand close together facing one another. They start to tremble at accelerating speed with sideways motions, as the rubbing of their locked bills produces a clattering sound.

Once the impassioned bill-in-bill business is over, the female ducks down a little and the birds (*Vögel*) proceed to "*vögeln*."[12] The ibis's shaking game begins at once; fluttering his wings, the male tries to keep his balance on the female's smooth back. At the same time, he grasps her bill and pulls her mane and head toward him; their two heads shake vigorously from side to side, producing more bill-clattering music accompanied by many calls of *chrup*, while the male wags his tail from side to side and then in circles. The female moves in a similar way, then lifts her tail sideways, whereupon he immediately sinks his tail downward. Wings outstretched, he presses his cloaca onto hers, then releases her bill and slides off her back backwards—and bows to her once again.

The postlude to mating is carried out with nice synchronism. Ornithologists call this the "post-copulation display."[13] Both birds straighten up with bills upraised, crests erect, and rattle their beaks, blink their nictitating membranes, pitter-patter sideways in a circle, or shake their iridescent feathers.

In a show of devotion, the ibises stroke each other from crest to toe, a procedure technically called "social preening to reinforce pairing." A bird's urge to preen is transmitted to another by body language. Bill shoved under its belly, crest stiffly upright, wings drooping, tail lowered, the bird to be preened goes into a slightly submissive position standing beside its desired preener, male or female. When its feathers have been freshly cleaned—every single one is pulled through the gentle tongs of the bill—and its legs and toes have been cleaned by tender nibbling, the bird gives an oh-so-pretty bow once again.

Young female Waldrapps are often extremely clumsy when first attempting to preen. Many a Mademoiselle Waldrapp trips lightly over toward the preenable bird, forgets to bow, badly messes up his beautiful feathers, and then takes off, pursued by one angry, disheveled bird. If he catches the disheveler, she has to face the attacks of his flexible bill. To appease his anger, Mademoiselle cowers, puffs herself up, and pulls in her neck. If that has no effect, she stands up straight as an arrow and tucks her naughty bill into her fluffed-up breast feathers, as into a case. That will calm her attacker, whose bill blows gradually segue into a bow of greeting and ultimately into preening.[14]

Bill fights between competing male Waldrapps have something in common with Chinese shadow boxing: the birds hardly touch their opponent. A *whiff* can be heard when a bill doesn't hit anything but air. During these ritualized fights, the birds raise their crests and alternate between threatening clatter and whiffing. This mostly happens when defending the nesting site or nesting material.

Unpaired males arriving from their winter quarters try to attract females with a kind of nest-building pantomime at their chosen breeding site. A slow closing of the nictitating membrane shows the degree of their excitement while they puff themselves up in a submissive display, spinning around and tentatively scraping with their feet. If the male is fortunate enough to find, bow to, bill, mate with, and preen a female, then nest construction may begin. They set to work together. First, the old nesting material is thrown over the cliff. The old nest was messy and dirty; the new one will be messy but clean. The bowing birds have no interest in constructing a beautifully woven nest. Twigs, fairly strong branches, and brushwood form the crude framework for the nest site, which is "whitewashed" over and over with excrement; bundles of dry weeds, fresh tufts of grass including roots, and earth are built into the frame. While a bird of either sex searches for construction material, its mate guards the nest. Twig thieves and breeding-site usurpers lurk around for an opportunity to serve themselves. Instead of laboriously carrying twigs through the air in their bills, the robbers pull the choicest pieces from the edges of their neighbor's nest just as he is trying to keep his balance on his mate's back. If the guardian of the nest is careless enough to vacate his seat, he must be prepared to find it occupied when he returns or to watch his neighbor drag the whole nest structure over to his own nesting site.

The bird that trooped off with other construction-material foragers is recognized from a great distance by the bird squatting on the nest; after he lands, she welcomes him with a nest-building pantomime. Upon returning her greeting, the bird carrying the material throws what he has gathered at the nesting bird's feet, or even over her head. If he has found especially fine pieces, he will have presented them to his opposite number beforehand with an impressive display, his head held high. Now the birds push at the material to make a crude wall, one shoving from the outside, the other in the center of the nest from the inside; then they "shake in" with their bills some finer material to give the nest the required stability. The Waldrapp's "shakability" is great. If they suddenly forget what they are supposed to do and shake the material in their bills off into nowhere instead of into the nest, then the game of making the beast with two backs will soon follow.

Once the nest is fairly stable, it must have a hollow. The Waldrapp uses its upper body to make one, with bill stretched out straight and head lying flat on its side on the nest floor. Squatting on its heels, it shoves its body back and forth, up and down, against the walls of the nest while scraping away with its feet.

Before the female lays her first egg from her twitching cloaca, a so-called false copulation takes place that reflects the start of the egg layer's domination of the nest. Now it's her turn to balance on the male's back by spreading her wings, bending his mane and head backwards in her bill, and synchronizing her bill-rattling with her tail movements. Contact with the cloaca usually doesn't take place, but the mutual bowing after the mock deed is done is never omitted.

The Waldrapp's pointed oval eggs are somewhat larger than a chicken's; the shell is sparsely sprinkled with a color varying from olive brown to bronze and is delicately marbled on a blue-green ground. As a rule, three to four eggs are laid within several days. The twenty-eight-day incubation period begins with the laying of the first egg.

After all the eggs are laid, the parents take turns incubating. During this time, the birds' face, throat, and legs are blood red. While one bird keeps the eggs warm by fluffing its feathers—turning the eggs gingerly and often with its bill—the other one flies off with a contingent from the colony to their feeding grounds, where they move in concert with measured strides in pursuit of prey. Having satisfied their hunger, they gather up some decoration for the nest. The business of incubating is tiring. The time there will pass more quickly with a large feather from their enemy, the Eagle Owl, or a small root formation polished by the river, or a scrap of paper, a colorful rag—all these will be "shaken" into the nest with the partner's help. After bowing in devotion to its mate, each returning bird proudly presents its gift and showers it upon the incubating partner.

Peeping sounds in the eggs announce the approaching end of incubation. Peeping begins when a little bird has pushed ahead with its bill into the lens-shaped air chamber at the flat end of the egg. During the hatching process, or pipping, the sound becomes a shrill squeaking that won't stop until the birth pangs are forgotten because of its parents' preening care and the refreshment from their bills flowing down its little throat.

It is difficult for the chick to pip all by itself. Hatching does not really involve something as simple as just "slipping out."[15] First, the chick has to rub the bump on the tip of its bill—its egg tooth, which contains a grain of sharp aragonite—against the shell so as to make some cracks. When the shell starts to break, the egg membrane does, too, and light finally bursts into the shell through some small gaps. When the chick stretches its legs, it pushes its head through the "roof"; its head is protected only by a cap of fluffy down. The shell fragments are hard and pointed, and the pipsqueak of a Waldrapp has to remove them; the membrane is tough and clings to its naked, amorphous belly and to its translucent pink skin covered sparsely with natal down. It takes one and a half to two days after breaking through the roof for the chick to be completely free of egg shell. Any remaining splinters or bits of membrane are nibbled at and picked off the exhausted newborn by its parents.

A few days after hatching, the slits on the nestling's bulging eyes open—it could peek through them on its very first day—freeing its blue-gray to yellow-gray irises surrounding brownish-bordered pupils. It will take two whole years to complete the gradual change from gray to orange.

By the second day, the nestling has recovered so well from the hard work of hatching that it can rehearse cooing softly during its parental preening and nibbling treatments. This song of delight will now always be sung whenever its feathers are cleaned. Whether a sibling or a parent does the nibbling or preening, its cooing won't stop until this enormous pleasure lulls it to sleep and the heavy head on its scrawny neck sinks onto the head of a neighboring nestling.

The transformation from chick fuzz to juvenile plumage goes relatively quickly. The casings, known as blood feathers, that emerge from the skin after one week break open a week later, allowing the iridescent, metallic green vanes of the black feathers to unfold; the wing feathers already have a touch of purplish sheen. The small facial and head plumage is gray with a white fringe. Gray, too, is the wrinkled skin around the chick's eyes; gray are its sturdy little legs. They don't turn coral-red like its bill until the twelfth week. A little crest of lancet-shaped feathers appears as early as week three.

As one squeaky chick after another pips, many young Waldrapp pairs that are experiencing the more-than-a-week-long hatching period for the first time start quarreling vigorously. Amid much threatening behavior, bill-clattering, bleating, scolding, and whiffing, they fight for the right to warm, feed, and preen the chicks. In the end, it's usually the male that feeds the nestlings in the first few weeks. He bends over the young bird, takes its briefly opened bill—which has just started decurving in the second week—in his own and forces and twitches a thin, predigested, nourishing mush down his chick's gullet.

A would-be beggar starts practicing early. A nestling's life depends on its ability to beg. As long as its perpetual peeping continues after it hatches, its parents show it special attention while feeding and preening. After that, the rule is this: the pushiest beggar gets served first. If there's a food shortage while the chicks are being reared and a late pipper can't keep up with the begging skills of its bigger siblings, it will get nothing and eventually starve to death. Without further ado, the parents will toss the dead thing out of the nest or "shake it" into the nest's structure.

The worst of the grueling hatching process is scarcely over before the nestling's begging lessons begin. But since it can't yet hold its head upright on its outstretched neck, its head wobbles back and forth alarmingly while the greedy little tyke begs with a *lib, lib, lib* or *yook yook* sound. In no time, its begging song swells to a respectable trill that ornithologists describe as a "chirrup whirring." A "hammer-like" nodding of its head reinforces the trill's urgency.[16]

The nestling has learned in the meantime to fill its own greedy gorge by serving itself from its feeding parent's bill. It draws its milky nictitating membrane over its eyes with relish, trembling and twitching as it swallows the more solid mush. It's already learned to pick out something from a crack here and there or to gobble up mush spattered around in the nest.

The older the Waldrapp offspring are, the more wildly they behave. Ultimately, they stage a sort of wild Indian begging dance, fluttering their wings and emitting crazy trills while tripping around their parents in circles. If their parents feel hard-pressed by their begging, they pretend to be asleep. If the trilling mob keeps it up, the parents have no option but to flee.

Soon the new generation will have learned a lot, apart from begging, by playing in the nest. The nest is the school of life. It takes just a month for the little ones to learn how to bow properly in greeting. The siblings automatically take to preening and plucking each other, and even their parents have to go along with it. Curiously exploring the nest up to its edge leads to practicing snapping the air and "shaking in" material into the nest, but with an empty bill. They also experiment with billing and retracting their necks. Wild chases precede playful fights, where the frolicking fledglings lunge and bite at one another in sport. At barely a month old, they begin to try out their wings, forming a standing crowd at the edge of the nest. Turning their backs to the abyss and fanning their tails, they flap their wings and emit hollow quacking sounds as they stare ahead and puff up their throats. Before their maiden flight, they flutter from jutting rock to jutting rock, still lowering their legs and stamping their feet in an attempt to take on the wind. They soon learn to use updrafts. The young wild ones will have mastered the air after about fifty days and will fly at one another. One will flip over onto its back and kick its playmate flying overhead with both feet, hitting it on the breast and belly. They never miss an opportunity to show God and the Waldrapp world how they can cut an audacious caper.

After the young aerobatic fliers have made an elegant landing on their breeding rock, some display ensues, where they lower their heads, stick out their necks, and spread and raise their tails; then they can turn to more thoughtful games. A few stones present a challenge: to roll them over the ledge and listen to them bouncing down the rocky wall before they plop into the Euphrates with a splash.

Fledged birds are fed and preened by their proud parents before and after each flight. When the young ones are on the feeding grounds making their first playful attempts at catching prey, the older birds proudly strut along ahead of them.

Picking off a land snail from a plant here and there is child's play. But picking insect larvae, scorpions, and beetles out of the ground takes more

Anita Albus, *Waldrapps in a World Landscape,*
oil on canvas, n.d.

skill; they have to learn how to probe the right spots. Greater hunting experience is needed to catch a mouse or lizard, which means racing after it while hugging the ground, neck outstretched, to catch it on the run.

Larvae in the shallow water near the banks of the Euphrates are welcome prey. This is also where the birds slake their thirst, which they do during their bathing ritual. Hugely puffed up, they stand breast-deep in water, shoveling water over their backs with their wings, first on one side, then the other. Now and then they give a shake, straighten out their feathers, and turning their heads sideways and making a gnawing motion to get a mouthful of the Euphrates, they let the water run from the scoop of their uplifted bill down into their throat.

A refreshing bath rids the young birds of the remains of their blood feathers; they shed most of the water by shaking their body and beating their wings. To dry their feathers, all ibises stretch their wings vertically over their back, bring them down again, and, with drooping primaries, abandon themselves to wind and sun.[17]

Bathing in the river is one thing, sunbathing quite another. Their plumage must be dry, the atmosphere calm, the day flooded with sunshine, the Waldrapp world at peace, and the semisacred bird "circumspect."[18] It stands on both legs, breast to the sun in the most erect position possible, and pushes a wing forward in slow motion with its "prow" parallel to the body, where it deposits or parks its wing like a shield. It stays that way for some time. Ornithologists call this overture to "sun worship" the "initiating posture," which culminates in the so-called Delta-wing posture.

The overture is followed by a fitful spreading of both wings, either stretched out horizontally with their inside turned toward the sun or laid down with the tips of the primaries touching the ground. After this pose, the completely symmetrical Delta-wing posture is assumed by angling the wings' "wrists" to bring both wing tips into contact—the Waldrapp, like an almighty bird god, raises its uplifted head on its long, extended

Anita Albus, *Sunbathing Waldrapp in Delta-wing Posture,*
watercolor and body color on paper, n.d.

neck over a generous "bowl" of wings proffered up to take in the light. Usually, its belly feathers are ruffled as are its rump feathers, invisible to the observer. Sometimes this spectacular posture is carried out, half-heartedly, as it were, with one wing; there is a whole array of potential variations and a high degree of complex movement sequences to the Waldrapp's sunbath.[19]

"Sun worship" can last up to an hour and a half *in extenso* and is often interrupted by preening. It's rather rare for pairs to engage in mutual preening during this time. Even if a whole "congregation" of Waldrapps gathers for a "sunning service," every member will stand apart by itself. If the temperature is high, the birds cool off by raising their back feathers and begin panting with bills open and a vibrating sound in their throats.

Almost all birds, even nocturnal owls, have a special sun ritual. Scientists have no idea why they do it. This much is certain: the underside of their feathers reflects thermal radiation in the infrared range; high heat makes ectoparasites, like mites and bird lice, move into the upper layers of the bird's feathers, where they are easier to pick off; heat makes the featherless part of the quill flexible so that feathers can be bent into shape and better organized; plumage coloration is protected by direct irradiation from sunlight, which prevents feathers from being washed out or turning blacker through progressive melanin retention.[20]

FROM ROCK LEDGE TO BREEDING BOX

That the Crested Ibis and the Waldrapp were identical was still unrecognized on February 16, 1879, when Charles G. Danford discovered, in Bireçik, a pair of Crested Ibises newly arrived from their winter quarters. A whole passel was gathered there two days later. Danford's article, "A further contribution to the ornithology of Asia Minor," informed the ornithological world of this colony.[21] Danford and two other ornithologists who came to Bireçik after him for the *kelaynaks* did not report exact data on the colony's size. According to municipal authorities in Bireçik, it was said that there were three thousand breeding pairs in 1890. The first halfway-reliable estimate comes from a fourth ornithologist, who went to Bireçik to observe the Waldrapp in the spring of 1911 and counted about a thousand birds.

A fifth bird observer arrived decades later to view the ibises and would dedicate a lifetime to studying them. On June 8, 1953, Hans Kumerloeve first stood on the eastern (Mesopotamian) bank of the Euphrates and observed the arrivals and departures of a band of Waldrapps nesting on the ledges and promontories and in the niches of the high rock face north of Bireçik, which was crowned by a citadel. The colony extended from below the citadel up to the surviving steep rock walls within Bireçik that separated the upper and lower towns. Kumerloeve tallied thirteen hundred Waldrapps "including the larger juveniles that were already visible,"[22] which added up to about four hundred breeding pairs.

The breeding site below the citadel was completely deserted when Kumerloeve went to Bireçik a second time, in 1962. The word in town was that the birds had long stopped nesting there because the soft limestone of the cliffs had been crumbling for years. The nesting sites were now restricted to a rock face in the center of town, where a colony of the remaining 250 birds carried on the business of breeding amid the infernal racket of the building frenzy that was transforming the *Bir-evçik* into a large city.

The flat roofs of the new buildings, only a stone's throw away from the nesting ledges, were ideal observation posts for an ornithologist, unless a sea of television antennas was blocking the view. There sat Kumerloeve, at the beginning of April and of June nearly every year, following the Waldrapp colony's tragic gradual decline. In 1964, he counted one hundred breeding pairs; in 1965, seventy to seventy-five; in 1967, forty-five to forty-eight; in 1968, forty-five to forty-six; and in 1969, on his last visit, the remaining seventy-eight birds made for thirty-eight to thirty-nine pairs.

It hadn't taken long for the protective, overhanging rock in the city to begin crumbling as well. Eggs would roll out of the nests from downward-sloping ledges. Construction material and garbage was dumped from the buildings above the nests, while nasty boys got a kick out of throwing stones at the funny black birds and watching the terrified, unfledged nestlings tumble over the precipice.

No talk now of Noah's messengers, about their gift of bringing the spring, their protection of pilgrims to Mecca, their role as soul-bearers. The last bird-welcoming festival was held in 1958. It was, at the same time, a farewell festival for the ferrymen who had always put on the ibis festival; since a bridge over the Euphrates was to be finished in 1959, the ferrymen would be needed no longer. But before the bridge was even completed, construction workers were stricken with bouts of malaria. To fight the epidemic, the Turkish health ministry deployed aircraft to spray huge quantities of DDT over the contaminated area. They began draining the swampy Euphrates lowlands, and after the intensive dousing with pesticides, not a mosquito, not a mouse was left. As if that weren't enough, swarms of locusts threatened to invade Turkey from the south— so why wait until they got there when you could make a preemptive strike with planes and the most potent insecticides?

The effects were disastrous. Anything that could wiggle or wing its way along was wiped out. Not even scorpions survived the poison attacks. People and pets fell ill. Insectivorous birds were hardest hit. Hundreds of dead Waldrapps were picked up inside the city limits of Bireçik alone. The reproductive rate of the surviving birds was affected for decades. Many eggs were so thin-shelled that they shattered; others lacked embryos. Based on the low hatch rate and the high number of nonviable nestlings, the colony's population pyramid shifted to the "urn" or "onion" shape typical of a predominantly older population.[23] Since Waldrapps in the wild do not breed until their fifth or sixth year, the number of breeding pairs kept dropping. Like the sixteenth-century Waldrapps doomed to become a rarity, it was now only a question of time before there would not be a single Waldrapp left in the former Crested Ibis's Land of Cockaigne, or land of contradiction.

Once things start going downhill, everything speeds up the process. What put the nail in the endangered Bireçik birds' coffin was, of all things, a Waldrapp protection program. The colony's inexorable decline attracted the attention of the World Wildlife Fund. A "project leader for the rescue of the Northern Bald Ibis in eastern Turkey" was appointed with much haste. Udo Hirsch was to start by determining the causes of the decline. He first came to Bireçik in 1971, to find thirty breeding pairs and eleven juveniles still there. A year later there were only twenty-three breeding pairs, which raised a total of six to eight young. Meanwhile, Hirsch saw it as his mission to educate the populace about reasons for protecting the *kelaynaks*. In particular, he targeted the numerous immigrants who came to Bireçik from areas where the Waldrapp was unknown. These people had never heard of the sacred nature of the birds and seemed not to be sufficiently alarmed if mean kids aimed their sling-shots at the ibises. Educating them was not quite as futile as the attempt to stop construction on the land above the breeding rocks, or as a request that became infamous in the eyes of the poor: to demolish the flat-topped buildings in close proximity to the nests. If the impoverished resident population had once been given food in celebration of the birds, these same people were now supposed to accept homelessness in order to save the birds. Holy *Kelaynak!* Only a Hirsch made gardener could think up a thing like that![24]

Even the belabored reintroduction of the mid-February festival to welcome the harbingers of spring had only the name in common with the ancient custom. Knocking a hole deeper into the rock behind the too-narrow breeding ledge proved to be an unsatisfactory attempt at overcoming the birds' lack of room. All that was achieved by extending the ledges with projecting planks was to deface the breeding rock. Fewer and fewer pairs came to Bireçik in February to breed; fewer and fewer juveniles made the long journey with their parents to their African winter habitat at the end of June. What happened to them there is unclear, whereas their decline in Bireçik is documented down to the last Waldrapp.

Since Waldrapps kept in the favorable conditions of a zoo would breed as early as their fourth or even third year, the World Wildlife Fund (WWF) experts hit upon the disastrous idea of compensating for the constantly falling reproduction rate by launching a program of breeding and reintroducing birds into the wild. There was no dearth of warnings, but the experts all turned a deaf ear and even the Turkish government finally stopped opposing the project. Kumerloeve, the great Waldrapp aficionado, did not find out about the project until it was too late:

When I arrived at noon on June 7, 1977, I counted only eight adults and eleven juvenile Waldrapps on the colony's wall, plus five to eight in flight. Arriving at the *fidanlik* [tree farm] on the bank of the Euphrates, I heard, much to my surprise, about a planned resettlement that would service two aviaries that were already built. I witnessed a short time later how three, barely two-thirds grown juv. were brought in and placed with three previously taken juv., how people fussed over their food or tried to stuff them with lamb's meat etc. Late afternoon on June 8 there were two ad. and six juv. on that part of the rock ledge and two more juv. on the wooden ledges there; on the morning of June 9 there were doubtlessly the same juv. birds of the year as well as eleven adults present; plus the six juv. in the aviary. When I returned to Bireçik on June 22, one aviary was occupied by 9 young (in satisfactory condition), the other one by two ad., whereas there was only a single adult, which either stayed on the breeding rock or flew around aimlessly. Making a final check on June 23, I could not see one single free-flying Waldrapp. Some or all of the ad. might have flown away. I could not reconcile the difference between the nine aviary-juv. in the last third of June and the fourteen juv. counted about two weeks earlier. We can only hope that those nine survived—however greatly uncertain and problematic their fate and the planned project may be in view of the fact that until now no one has had any experience with reintroducing free-flying Waldrapps. And the Bireçik population is now much too precious and unfortunately too thin for taking an unpredictable risk (they are thinking of a rock face for them on the same bank of the Euphrates ca. three-four kilometers [two to three miles] farther north). On the other hand, we can say very little about whether and how long the small remaining flocks might be able to hold steady. The next few years will tell, one way or the other—but certainly the painful feelings of those who can never forget the former spectacular breeding colony cannot be swept away.[25]

The five juveniles and single adult that Kumerloeve found missing had died while they were being captured. True to the motto, "Only Turkish food for Turkish birds: we have no interest in tried and true recipes from European zoos," the first chicks incubated in the nursery aviary were stuffed with indigestible lamb's meat and whole grain. Only after the chicks had died a most miserable death did their Turkish caretakers learn their lesson. But money for enough expensive food was tight, so the aviary chicks wasted away "or flew off with physical defects, like crooked bills, drooping wings, or crippled feet."[26]

Construction work began in 1977 on the Waldrapp station three miles north of Bireçik. The birds were to be kept in two large breeding pens supplied with feeding stations and were to multiply diligently; moreover, they were to lure over the wild population on the wind-blown breeding rocks in Bireçik. A wooden shelf for nesting on the naked rock face inside the station was to facilitate undisturbed incubating for the thus enticed *kelaynaks*.

Everything was ready to go at the beginning of March 1978. The captive birds were taken from the small aviaries to the large ones in the newly built breeding station, and another adult was captured from the colony that now totaled thirteen pairs and four unpaired adults. Of the twenty-nine young hatched in the city, only sixteen were to fledge. Ten of these were caught and placed in the aviary with three adults. This time the process of capturing them cost "only" three juveniles their lives.

There were still nine Waldrapp pairs nesting on the rocks within the city in 1979, and three other pairs were lured over to the wooden brooding platform in the breeding station. Grass and tamarisk twigs from the tree farm, for nest-building, were scattered with extreme care in the station yard. But no one figured on the wind that swept along the smooth rock wall. The light grass, the delicately woven tamarisk twigs were blown off the platform in no time. No matter how hard the birds tried, they just could not manage to enlarge their nests. Ultimately, ibis instincts triumphed. The birds rejected the material so eminently suitable for "shaking in" and flew in proper nesting material from a long way away.

The Waldrapps knew how to help themselves when nest-building, but their chicks were helplessly abandoned to the fierce sun shining on their incubation site, which had no shade the moment the sheltering bird was disturbed and flew off the nest. From noon on, the sun beat down on the breeding ledge. Even in March, the adults panted heavily in their nests. In the first days after hatching, the chicks were constantly sheltered, but the little ones didn't learn to pant until two weeks later.

Five young birds from the three wild Waldrapp pairs avoided dying from the heat; the nine pairs in the city center fledged sixteen young; in the aviary, four pairs laid twelve eggs, producing ten chicks, all of which died in their first few days. Another two adults and five juveniles were captured for the aviary that year. In August 1979, eight adults and twenty-three juveniles were living in the two aviaries in the station.

The aviary bird community attracted more and more wild birds to the station. In 1980, fourteen of the twenty-three Waldrapps returning to Bireçik went to that confounded breeding platform above the aviaries. Two pairs fledged two or three young there; the rest of the adults were scared off by attempts to capture them and moved back to the breeding rock in the city, where six pairs were able to fledge eleven young. Eight pairs were formed from the twenty-nine birds in the aviary. Seven pairs bred successfully. Seven of eighteen hatched nestlings died. The eleven survivors did fledge but suffered so much from rickets that a release was ruled out. Hirsch's program actually foresaw integrating fledged captive birds into the wild population while they were still in their hatch year.

Failed breeding preceded failed reintroductions into the wild. The first attempts were made toward the end of the 1981 and 1982 breeding seasons. Several birds died, and not a single captive Waldrapp migrated with the rest of the colony to their wintering grounds.

Hirsch blamed the late release date instead of the fact that the birds were not adequately prepared, a miscalculation that was to pave the way for the disaster of March, 1984:

> Sixty-seven birds were released in total. Only twelve migrated with the colony to their wintering grounds, and these were probably all birds taken from the wild. Juveniles from released birds migrated without their parents. These young birds were surely at a disadvantage because they have much to learn from their parents at this stage in their lives. The released birds had not been prepared for their reintroduction (for example, they could have been offered living prey). They mobbed the wardens who fed them every evening, just like domestic chickens. Only a few former aviary birds survived the hard winter in Bireçik; only two of them were left by 1989. Reintroductions of subadult and adult Waldrapps have not been crowned with success. This has also been demonstrated in Israel where many attempts were made using different methods.[27]

Only four Waldrapps came back from Africa to Bireçik in 1988. In 1989, there were but three. One of these fell victim to high-voltage wires. After the last wild Bireçik ibis had waited in vain for a female, in spite of frequently repeating his nest-building pantomime, and the thirty-nine birds released from the aviary kept circling over the empty station, even a

"Hirsch" couldn't fail to notice that the Waldrapp conservation program had collapsed. A Waldrapp postage stamp and a Hotel Kelaynak were all that remained in Turkey of Noah's messengers.

MOROCCO AND THE MAREMMA

Two of the formerly thirty-eight breeding sites in Morocco are the last refuges of the Waldrapp in the wild. Both colonies nest on sandstone cliffs of the Atlantic coast, one north of Agadir, one south.

In 1996, forty birds died of unknown causes within nine days, after which the population rose in the last three years from 220 to three hundred Waldrapps.

A rump colony of seven birds was discovered in northeastern Syria in the spring of 2002. It had escaped scientists' notice that *Geronticus eremita* had been dispersed throughout a remote area in the Palmyrene desert until two to three decades before. The seven adults that had not yet wound up in some poor Bedouin's pot flew off in mid-July 2002 with their three fledged young to their unknown wintering grounds. In February 2003, six of them returned to the breeding rocks in the Syrian Desert steppe, but without their three "teenaged" young. Bedouins were hired to guard and observe the Waldrapps.

Thanks to this defense against egg thieves and hunters, three pairs were able to migrate to their wintering grounds in July 2003, with seven fledged young. As in the song of the ten little Indians, only five adults returned to the Palmyrene desert in February 2004. Apparently there are gourmets in the birds' wintering habitat with a fondness for tender, young Waldrapp flesh.

As difficult as it is for the Waldrapps to hold their own in the wild, it is easy for the tame ones to multiply. Although the former are nearly extinct, many zoos can't cope with all the latter ones. There are more than two thousand Waldrapps worldwide. Except for the few migratory birds left in Turkey, all are descended from the Moroccan Crested Ibis.

This is a resident on the Atlantic coast that is genetically different from the Turkish migratory birds.[28] But the ancestors of the zoo population have almost all been taken inland from Middle Atlas breeding colonies, which have been gone for decades, just as the colonies in the High and Anti-Atlas ranges have disappeared. The Waldrapps that bred in the Atlas Mountains were visitants. By the time snow covered the slopes at the end of October, they had flown to the lowlands west of the mountains. If conditions were right, they would visit as far as the coast, where they were more or less welcomed by their resident relatives.

The reintroduction of caged *kelaynaks* in Turkey failed for the same reasons that the attempt to resettle *Geronticus eremita* to Israel failed. Migrant birds' legendary powers of orientation will die in most species when they are raised in captivity. Captivity eradicates their gift of travel. They still have a magnetic compass, to be sure, but because they haven't learned to tell the compass points by the sun, the stars, and the polarization patterns of sunlight, they cannot calibrate their compass. All that's left of them is the restlessness felt at migration time. Without the guidance of experienced parents, young Waldrapps in Turkey and Israel either wandered around disoriented or disappeared, never to be seen again.[29]

The experiments in Birecik and Tel Aviv had come to naught, but then the ethologist Ellen Thaler and her former doctoral student in Innsbruck, Karin Pegoraro, thought up a procedure tailor-made for releasing Waldrapps. They had read their Gesner: If the young are removed from the nest some time before they are fledged, they may be easily reared and tamed "so that they fly out to the fields and quickly return."

It was even easier to tame them *ab ovo*. Just as a Waldrapp raises young in the wild, their upbringing in captivity begins with someone talking to them through their shell. "It's okay, I'm here!"—this is how the foster mother calms her chick when it begins to peep inside the egg in an incubating box. Pegoraro raised two Waldrapp chicks for observation

Otto Kleinschmidt, *Pair of Waldrapps*, 1897–1905

as part of her doctoral work on *Geronticus eremita* as early as 1985. Now six little ones were hatched out of their shells in the incubating box. The chicks looked upon the bill-less, pinionless, and featherless giant women helping to liberate them from their enclosure as belonging to their own kind. The substitute parents, Pegoraro and her student Susanne Stabinger, were the only people the birds tolerated; they would not adapt to anyone else. The six Waldrapp offspring were named after the colors on their bands: Turquoise, Blue, Alu(minum), Black, Yellow, and Gold. They were fed first on crickets, grasshoppers, and baby mice, then on snails, grubs, beetles, and dew worms so that they developed in the desired manner. Their "nest" was placed in the rafters of a farmer's cottage with an open gabled front. They could be encouraged to fly after about forty days. They maintained steady eye contact with their flightless parents in the garden, circled the roof ridge, and were heaped with praise after they landed in the garden. A Red-billed Chough that had grown up with them taught the fledglings the dangers and enemies to look out for. Nobody had to instruct them how to probe in the ground for insects and larvae. Nonetheless, the morale of the Waldrapp required the foster parents to keep stuffing the birds' voracious gullets for four more months.

Turquoise, the first hatchling, was particularly fond of Pegoraro. One evening, after she had been absent for three days, she discovered that the bird was missing from the attic. Staubinger found Turquoise under a high-voltage tower, dead. When two more birds went missing in November, it was decided to go back to the aviary. The countryside was white when Black, Yellow, and Gold returned to where they had come from seven months before: the large Waldrapp aviary in Innsbruck's Alpine Zoo.

The successful reintroduction of the Innsbruck Waldrapps was a model to be imitated. One day, not too long after, Waldrapps raised with the same close-to-nature methods were reintroduced from the zoo in Jerez de la Frontera into the Sierra El Retin in the southwest of the province of Cádiz. The region meets all the conditions worthy of a

Waldrapp's existence: impenetrable rocky cliffs for nesting, open country with a climate mild enough to provide food in the winter months, a river for drinking water and bathing, and an area that is off limits to mischief-makers—it is the site of a Spanish naval instructional facility.[30]

Thirteen years have gone by since the Innsbruck project, and not only have the zoo Waldrapps multiplied but their Noahs have as well. It's questionable whether the trial of reintroducing the birds into Austria and taking them over the Alps to the Maremma guided by ultralight aircraft will succeed. The birds are simply not migrants anymore.

We have much to learn about Waldrapp behavior from this incredibly expensive project, which is at the same time, of course, good advertising for its company sponsors. In August 2003, hand-reared Waldrapps followed along, as visitants should, behind their helmeted foster mothers in the ultralight in stages; they were guided in flight over a migration route across the Alps. But why should they keep flying when they saw land and the promise of prey below? Again and again, they had to be recaptured and driven ahead by car. Finally, after flying as far as Jesolo, they were taken in several stages, at times in a large mobile aviary, to the southern part of the Maremma, where they spent the winter with their adoptive parents. There is no question that after the aviary door was opened in the morning and they circled over the Laguna di Orbetello those twenty-one birds were the happiest captive-bred Waldrapps on earth.[31]

CHAPTER 6

The Shy Corncrake

NAMES ARE only names—that's the first principle of scientific nomenclature. But Linnaeus took the Corncrake's scientific name from listening to the bird. *Crex crex* is both name and call. The majority of the bird's folk names imitate its sound. If they're not based entirely on the rail's *Arpschnarp* call, almost all of them allude to its rasping, snoring, growling, creaking, grating, scratching, rattling, or chattering. Even if we knew nothing about this "rasping chick" (*Schnarrhühnchen*), we could still learn much about its nature from its 118 German dialect names. It lives like a quail, hidden in high grass and among other plants. The Corncrake was regarded as "the quail king" because from a distance it is reminiscent of a rather large quail.[1] It also bobs its head gracefully with every step, just like a quail. It is a loner, as befits a king. There is a belief as ancient as Aristotle and Pliny that the Corncrake leads the quail nation when migrating from their wintering grounds because it returns as late as they do. *Ortygometra*, "quail mother," was its Greek designation. Its vegetational habitat is reflected in the names "broom rail," "barley rat," and "little grain hen" as well as in its numerous grass and meadow appellations. We learn from the names "mowing bird," "scythe-whetter," and "mowing witch"

Pierre Belon, *Corncrake (Crex crex)*, 1555

about the sad fate of many a female Corncrake, or "lazy maidservant" of a "quail farmhand" that was beheaded, spiked, or cut in two because she was incubating at mowing time. The name "night crier" teaches us that the crake's call, which is like the noise a comb makes when you rub a stick back and forth over its teeth, is heard especially in the dark. We suspect from the names "grass runner" and "meadow runner" that it is nimble of foot—"It seems to shoot away over the ground," as Naumann puts it.[2] It seems like a scurrying "phantom" that "resembles a rat more than a bird."[3] It pops up as a "gray Punch" when it stretches its bluish, ash-gray neck vertically to its full length and spies all around; from head-on, it looks like a hand puppet with its feathers held smooth and flat against its sides, like a rail, and after drawing in its head it vanishes into the grass just as suddenly as it had darted out. The name "black clown" applies only in the dark; otherwise only little "quail princes and princesses" are dressed in a black coat of down.

Just how mysterious the Corncrake's existence—as audible as it is invisible—can be emerges from a few pages in August Strindberg's *A*

Blue Book. "The Miracle of the Corncrake" is a conversation between a teacher and his pupil one summer's evening during a stroll through fields of clover:

> ... and they heard a sound, something like "crex, crex."
>
> "What's that?" the teacher asked.
>
> "A Corncrake, of course."
>
> "Did you see a Corncrake?"
>
> "No!"
>
> "Do you know anybody who has seen it?"
>
> "No!"
>
> "Then how do you know it's a crake?"
>
> "It's what people say."
>
> "Look here! If I toss a stone at it, will it fly?"
>
> "No, because it can't fly or is a poor flier."
>
> "But it migrates to Italy in the fall. How does it do that?"
>
> "I don't know."
>
> "What do the zoologists say?"
>
> "Nothing."
>
> "Do you think it flies over Öresund, goes through Germany, migrates over the Alps, or goes through the St. Gotthard tunnel?"
>
> "They don't say."
>
> "Well, now, Brehm figures there's a pair of larks for every acre of field and meadow. If we calculate one pair of crakes to every hectare, then there are five million in our country in the spring. And after it lays seven to twelve eggs in a summer, that makes thirty-five million Corncrakes in our country in the fall. Wouldn't you see them when they fly or migrate over Öresund?"
>
> "I can't explain that."
>
> "A poor flier can't fly over Öresund—could they possibly walk around the Gulf of Bothnia?"

"No, because it has to go across rivers, and you'd see a parade of them, like lemmings. Besides, there are seventy million crakes in England every fall, and they can't go by land."

"Then is it a miracle?"

"What's a miracle?"

"Something you can't explain but have no right to deny."

"Then the Corncrake's migration is a miracle and must be following unknown natural laws, or else it's supernatural."[4]

The riddle of the Corncrake's migration would not leave Strindberg in peace. Later on in *A Blue Book*, he has the pupil meditating on the "birds' secrets":

The Corncrake still hasn't told us its secret, how it migrates. The Irish believe it's a resident, though they don't know where it hides. The English thought it was a ventriloquist that could throw its voice a long way away. That's because they heard it where it couldn't be—they wouldn't see a blade of grass moving where it was supposed to have run away. Since it's a very poor flier, a flutterer, actually (so Brehm says), it can't fly to Africa. Brehm thought it was a migrant, but he didn't think of the English Channel, Öresund, or the Alps. And so it must have stayed. Then the question is this: where does it hide in winter? Some guess in hollow trees, ruins, and what not. But you might expect it's Mother Earth, of course, who hides away all sorts of snakes, frogs, rats, mice, snails, and so on in the wintertime. A garden snail can dig half a meter down. The kingfisher digs its nest hole a meter underground. Why can't the Corncrake dig itself in, too, or use a molehill for its lair in winter? I don't see any reason why not!

A natural scientist puts seeing before believing. That's all well and good. But no scientist has ever seen a Corncrake in migration, so how can he believe it?[5]

Anita Albus, *Corncrake*, oil on vellum, n.d.

"No one has ever seen a bird in hibernation," the teacher might have argued, because in Strindberg's day only the Hopi Indians knew that the Common Poorwill (*Phalaenoptilus nuttallii*) slept through the winter. There was a folk belief in England that "corn rattlers" would turn into rats for the winter, and the Tatars questioned by Johann Georg Gmelin on his Siberian expedition assured him that the "meadow runner," which could barely fly, was borne on the backs of migrating storks.[6] Zoologists puzzled for a long time over the question of how such a weak flier could manage to migrate to southern Africa. Of course, they could have read their Buffon and seen that the "broom rail" departs by night and knows how to take advantage of favorable winds during its long journey. But even Brehm's fourth revised edition of 1911 still states that the Corncrake's nocturnal migration is probably largely done on foot. The Corncrake's exact route through the sub-Saharan savannahs is not known even today. People in various countries have put transmitters in the birds' plumage in order to explore the hidden life of *Crex crex*, but no one has yet tracked a migrating crake this way.

It's rare to flush a "meadow rasper." If you do, it will flutter off, keeping low over the vegetation with legs dangling, and drop back down after a few yards into the grass, where it sticks its neck out as if in flight and tries to shake off its pursuers by running away. It is after all a friend to secrecy and abhors exposing itself to its enemies. In the open air it is not in its element, but under a leafy roof of grasses and flower stems, it is. Not a panicle stirs when the "phantom" scurries along its secret passageways through the lea. Because of the suppleness of its slim rail's body—its vertebrae are not grown together, or connate—it can slip between stems and stalks without setting them in motion.

"The birds' light cavalry" is what Linnaeus called the flight of birds, their gift of being borne on the air. Even Corncrakes can be blown by the wind over deserts and seas, though their labored flight is more like the "heavy" cavalry. One year, when the south winds were exceptionally strong, Poland's *derkaczy* returned earlier than was normal.[7] With its tarsi sticking out behind, the "longlegs" comes flying with slow, steady wing beats from its winter quarters.

A shrill rasping at the beginning of May reveals that the Corncrake has returned from Africa to the area where he hatched. Sunny, heavily flowering brome-grass meadows, sedge meadows, and open high-bush fields are his favorite breeding grounds, where he will easily claim twenty acres for itself, maybe even fifty or more. The kingdom that this "quail king" dwells in with his "queen" of the moment is only a part of this area, and he rules it only for as long as his eight- to ten-day courtship and marriage last. This is a time when he does not take well to other males coming onto his territory, and woe to them if they dare try it at night! Territorial defense is not achieved with mere territorial calls. If a rival crosses the border of his territory, our Corncrake stands erect, lifts his wings, and raises a great hue and cry with some hoarse screeching. If this doesn't intimidate his rival, he fans his wings downward and backward, keeping them near his flanks, then raises them until the tips of his primaries touch above his short bill, and he begins to growl with bill closed. He tries to appear even more threatening to his opponent when he goes on the attack: neck thrust forward, head lowered almost to the ground, he raises his breast feathers and turns his outspread wings around in a flash so that for seconds the upper wing points upward and the leading edge downwards. If he loses the battle, he must look for a new "kingdom"; if he succeeds in driving off his rival by pecking it with his bill and hitting it with the bend of his wings, he intones a triumphant *rerrp rerrp*.

Brehm notes that the Corncrake's mind turns to reproduction immediately after he comes back to his kingdom.[8] Before he reaches the point of mating, "Arpschnarp" often emits cries through the night with such persistence that Naumann wondered how the bird "can stand its nocturnal creaking without getting hoarse."[9] Because he constantly twists and turns his upraised head as he calls, his *rerrp rerrp* seems to come from different directions and fools the listener. He calls from out-of-sight places in willow brush, rushy areas, or meadows with high sedge and yellow flag, cuckoo flower, meadowsweet, canary grass, and horsetails. Since female Corncrakes apparently like comb concerts, the males form groups of usually three to ten comb-raspers that have a rasping contest, keeping their distance from the females to about 150 to 650 feet. Two high points mark their nocturnal comb-music[10] during the mating season. The hours before midnight and between three and four find the birds' vocal intensity at its height: a *Musikant* can rasp up to ninety times a minute.[11] But his courtship repertoire includes gurgling and whinnying sounds as well. Before bewitching her, the male would have already selected several nesting sites concealed in grassy thickets to which he can lure her. A scraped-out hollow, a few dried stems and blades of grass, some moss—all indicate a nest that he wants to finish constructing with her. It's up to her to show which site takes her fancy.

How the "meadow snorer" performs his courting dance around the female was a secret he long kept to himself. The man who revealed it is regarded as one of the fathers of comparative behavioral research. For his four-volume work, *Die Vögel Mitteleuropas*, Oskar Heinroth, an assistant director at the Berlin Zoologischer Garten, and his wife, Magdalena, reared all kinds of birds in the home the zoo provided them with: from Goldcrest to Crane, from Eagle Owl to Nightjar, from White-tailed Eagle to Kingfisher. Heinroth photographed and described all the birds' developmental stages and their behavior in each one, from egg to breeding plumage. After the Heinroths obtained a pair of Corncrakes, nine wide-awake chicks were to hatch from the female's twelve spotted eggs. Seven of the hand-reared "quail princes and princesses" were given to the zoo; a pair was kept back for a year: "The first hatchling was called Jeremiah because of the day on which it was born [on the feast day of St. Jeremiah]; it was a female, originally mistaken for a male because of its size. The youngest chick, that was to become an especially sturdy piece of work, was initially a small one, so it was called Lilliput..." Lilliput's

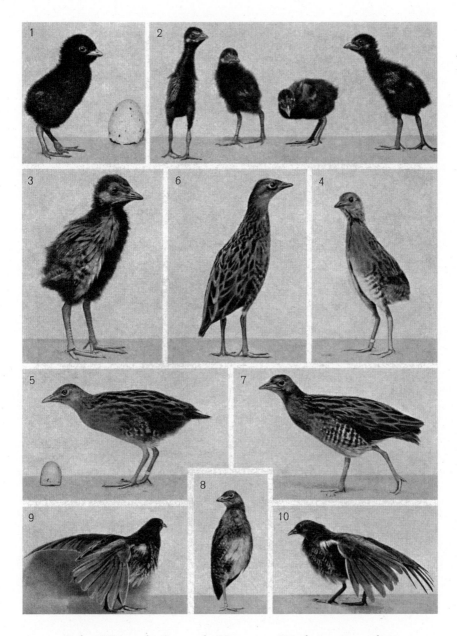

Oskar Heinroth, Corncrake No. 1: 1. at $1^1/_2$ days; 2. at 12 days;
3. at 15 days; 4. at 20 days; 5. at 30 days; 6. to 10. at 50 days, first juvenile
plumage; 9. and 10. sunning; vol. 3, 1926–33, photographs

Oskar Heinroth, Corncrake No. 2: 1. at 50 days, adult; 2. to 11. year-old
male in breeding plumage; 4. to 6. courtship display and, 7. to 8. feeding; 9. stepping
on a hand; 10. shaking itself; 11. ending of call; vol. 3, 1926–33, photographs

love life started up in mid-April 1926. In his fresh breeding plumage, he began to woo Magdalena Heinroth née Wiebe and acknowledged her hand to be his "quail queen":

He wasn't the least bit interested in his sister. Soon he started calling, indefatigably. The *rerrp rerrps* usually came in pairs that were repeated about two dozen times. The sound was enormously loud and hard, and since it started up out of the blue, any unsuspecting visitor was terrified by it. At the peak of his reproductive period, Lilliput would call hundreds of times, one call after the other, with only brief interruptions... Frequent loud drumming with a grunting overture could be heard, or else dull, hollow sounds. All we heard from the female was a *kyuck* that signaled anxious agitation.

Lilliput's love affair with my wife took the following course. Particularly when she held her hand out to him, he assumed a proud pose, spread his wings, and preened himself with his bill on his belly and flanks: in the bird world, preening movements are often related to displays of excitement and especially affection... Lilliput often pecked around on her fingers, and later ran over to his feeding bowl, fetched a scrap of meat, and presented it to her fingers... Sometimes it came to an actual courtship display. Then the bird would extend his wings particularly far backward, showing his cinnamon/chestnut-brown coverts, and would walk with mannered, broad strides around that hand... Lilliput never *crexed* during all these goings-on but only produced dull, groaning noises. If the back of her hand was held out to him, he would climb up on it, gently squeeze some skin in his bill, and then it would come to a proper mating on his part.

At the beginning of May we brought back one of his brothers we had given away in the fall in order to see how Lilliput would behave toward him. Oddly enough, the two males wouldn't have

Oskar Heinroth, *Corncrake: 4. at 1¹/₂ days;*
5. at 50 days, juvenile plumage; 6. adult male, courtship display;
vol. 3, 1926–33, colored photographs

anything to do with each other. When the new arrival *crexed* loudly the next morning, Lilliput was scared out of his wits, so we got the idea that he didn't know that he, too, emitted the same sound; it was out of the question that the strange male's calls might have stimulated some anger or jealousy. Since nothing was any different during the next few days, except that Lilliput got used to the newcomer's rasping, we took him away again. To make our ward more inclined toward his sister, my wife subsequently avoided being nice to him. We gave the pair a nesting site in the form of a box open on one side and with straw inside; Lilliput soon accepted it by digging a hollow in the straw. One day we discovered an egg there and hadn't noticed that Jeremiah, too, was caring for the nest. Several more eggs soon came along that both parents tended ... We found that the first eggs were fertilized. We hadn't seen any copulation, mainly because Lilliput kept choosing the human hand over his sister and had no interest in Jeremiah if somebody was in the room.[12]

Even if Lilliput hadn't been imprinted on Magdalena's hand, he still wouldn't have shown the slightest interest in his "lazy maidservant" after mating, in any meadow at all. With muffled gurgling and sharp whinnies, the wild bird circles its mate during his mating display, head lowered, puffing up his neck feathers now and then, and holding his outspread, bent wings to the ground. Her consent sounds like a donkey's *hee-haw* or the cackling of nervous geese.[13] Because the way to the heart lies through the stomach, he offers her, in courtship, butterfly larvae and other delectations before he mounts her again and again. In between copulations, they preen one another's beautiful feathers and do not let each other out of their sight. Once the planned nest has been finished together, there follows a tender tête-à-tête with grumbling and mewing—Bechstein says that the pair purrs almost like cats.[14] But the first or second of the eight to twelve eggs are barely in the nest when the "old farmhand" must look around for another female. His calls during his brief marriage, if he sounded them at all, were heard only by day. He has to show once again how persistently he can *crex*, now at night. Other "snoring quails" are not lazy either. Like his abandoned mate, the females have taken care of incubating, brooding, and raising young all by themselves. Scarcely two weeks will have passed since the mother took her first stroll with her black, velvety chicks before they must fend for themselves, because their mother is wandering off to greener pastures and another husband. And he, too, will leave during the laying period. She herself determines the time they will separate. The sound she uses to let him know that she doesn't want him around anymore sounds like a squeaky door.[15] So if conditions are right, he can go a-courting once again and maybe find a third wife. "Successive polygamy" is what ornithologists call it.

A Corncrake lays two eggs in a day and a half. But because all the chicks break out of the egg on the same day, the female begins to incubate her clutch with the second-last or final egg. It takes about seventeen days for them to hatch. Occasionally, the brooding female will cover up the clutch with greenery if she leaves the nest for a while to satisfy her hunger. Naumann and Brehm note that the female broods so assiduously for three weeks that she allows a carefully approaching hand to take hold of her on the nest during this time; she doesn't even fear the rustle of a scythe and will therefore sometimes fall victim to her faithfulness.[16] The most recent reports from the meadow tell us that the avid incubator will indeed flee in the face of danger and will seek cover as close as possible to the nest. Toward the end of incubation or during the hatching-out period, she might even attack a person who comes near her nest.[17]

The young spend the first twenty-four hours of their life whispering in the nest. Their down coats of silky, velvety black with reddish bronze iridescence dries out beneath their mother's brooding cover. The next day or the day after sees their mother already taking them out of the nest. Soon they are running away from her like quicksilver and returning with a whispered *eeyu eeyu* to her because they're scared at their own boldness. They "often gather together under her wings, scatter when surprised, scurry away over the ground like mice, and conceal themselves so adeptly that it is hard to locate them."[18] Should an enemy break into their meadow idyll, the female will try to lure it away from her chicks with shrill shrieks and a drooping, supposedly injured wing. If that doesn't work, she will, if need be, go so far as to attack even a dog.[19]

Pseeyu pseeyu is the chicks' begging trill. Although they can peck for food themselves as early as their second day, their mother stuffs their bills for three or four days. She shows them afterward what food is where

John Gerrard Keulemans, *Corncrakes,*
female (above) and male with chicks, 1897–1905

and how to get it: beetles are the preferred insect, followed by snails and dew worms. A skilled ventriloquist, she draws her gang, moving like weasels, to the feeding ground with a growling ventral sound and then, with a sharp knocking noise, invites them to peck.[20]

When the young, at just two weeks, must find their way alone in the meadow jungle, they still cannot fly and may have to run for their lives. Three more weeks pass until they are fledged. They spend their first nights as orphans huddled together for warmth. As soon as the vanes of their feathers cover their down, each little Corncrake goes its own way. If there are storks around, the chicks may not escape them. Otters, foxes, ermines, and polecats would also swallow up the tasty morsels if they didn't know how to hide so well.

Corncrakes' secretive lifestyles have protected them quite well from predators. While they might have had a chance to escape the reaper and his sickle or scythe in the past, nowadays there is no way to escape the farmer and his mechanical mower. Depending on the weather, the farmer will mow the meadow when the female Corncrake is either still brooding or leading her nestlings around. A fast, circular mower will get things done fast—who knows how long the weather's going to last? So the farmer starts on the outside of the meadow and circles in toward the middle, driving the birds seeking cover into the last island of grass, where they are finally caught in the machinery and will ultimately provide a feast for Carrion Crows. High up in the tractor towing the monster mower, the farmer can neither hear nor see the drama unfolding below. The mother bird's high-pitched cries of alarm and her chicks' shrill distress calls are drowned out by the raucous machine. The farmer probably still knows what his forebears called the Corncrake, but it's too much for anyone to ask that he see himself in the bird the way they did. He's certainly not a *Heckschär*, a "hedge clipper," anymore because with land consolidation there are no more hedges. The farmer's "field management" complies with a policy of subsidies, which rewards setting land aside but not cultivating hedges and grass fields or using crake-friendly mowing methods. The world of his farming forefathers has hardly anything in common with his own. Owing to the mass production of agricultural products, there are no more famines; but to exist the way the ornithologically minded naturalist-farmers Johann Andreas and Johann Friedrich Naumann did is unthinkable today—witness a letter from the latter. In answer to a request from Amsterdam about whether he would be willing to paint the plates for Temminck's *Manuel d'ornithologie*, Johann Friedrich Naumann wrote, on August 28, 1816:

I own a small farm, which supports me and my family but neither allows us to live in great style nor leaves anything over beyond the bare necessities of life. I must attend to anything that happens, and unfortunately often lend a hand myself; the income does not permit me to keep regular servants. My garden, in which I have a nursery and cultivate almost 700 species of exotic plants (I am also a botanist), I must look after largely myself, and therefore I am also a mechanic and make all the smaller tools for farming and gardening, and also my guns and other items of wood, bone, and metal. You therefore find me always occupied; now I am a carpenter, now a locksmith and gunsmith or turner, or I work as a gardener in my garden or I oversee my farmhands in the fields. To recuperate I go for a while into my little collection room and cheer myself up among my pets, or I hunt or catch birds. I am a good shot, and hunting is my greatest pleasure, because then the practical side of ornithology and in addition botany and entomology can be employed. That with so many occupations I have no time left to work for anyone else, you will readily understand, because though many jobs do in fact cease as soon as the bird migrations end in the fall, they are replaced by new ones more suitable to the season, which last until the spring migrations; for now I begin to organize the notes that I collected in the summer and to do my literary work in the evening; during the day I make paintings and copper-plate engravings of my birds. My elderly father cleans and polishes the copperplates, and he also prints from them during the winter, having also built the presses himself. So therefore, when in winter the farm and garden work is over, the jobs for our book, such as drawing, painting, engraving, printing, [and] describing the birds are our welcome occupations. Only the printing of the text and coloring of the plates we do not do ourselves; the latter is copied from sample sheets that I have made. Then the winter is also the time when I can study books on natural history [and] travel, and the latest periodicals, at my leisure.[21]

Farm consolidation, channelization of our rivers, transforma-tion of damp meadows into drained farmland, wetland destruction, and intensified hay and silage production—also called utilization of green space—have driven *Crex crex* to the brink of extinction in West-ern Europe. The fenlands have disappeared along with it—they were the Corncrake's own world, where blades of grass, with no knots, would stand up straight over its nest and pathways even after thunderstorms. Our cherished livestock despise hard grasses, which is why sedges are called sour grass as opposed to the sweet grasses in hay meadows. Even moor grass, which has no knots[22] and was once used to clean tobacco pipes and tie up grapevines, is not a particularly popular feed plant and gets used as litter, like the sour grasses and shrubs in bog meadows and all the plants in high sedge marshes in siltation zones of rivers and lakes. But meadows that are just for litter aren't lucrative. In the last fifty years they have shrunk to a fraction of their original size. Moor-grass meadows and fenlands are on Germany's Red List of endangered biotopes as areas threatened by complete destruction.[23] Whatever can't be "amended" by draining and fertilizing stays unused. In the long run, fallow meadows are not much help to the Corncrake. Soon they will revert to what they were before man dreamed up litter meadows that can't tolerate fertilizer and are mowed just once a year and as late as possible: they turn into bush and woodland.

Great expanses of meadow and pasture have only existed since the eighteenth century, when fences started to keep livestock out of the woods. But Corncrakes existed before the New Stone Age in a world as yet unshaped by man. Where did the crakes live when forests stretched from the Alps to the North Sea and the Baltic, from the west of France to Russia?[24] Biologists surmise that Corncrakes lived mostly in lowland river meadows kept open by ice shear.[25]

The Corncrake stays around civilization, and it is flexible as to its choice of habitat. Even dry meadows would be to its liking as long as they weren't mowed too early. Being shielded from sight and having few obstacles to walking are what it wants from meadow and field. If these conditions are satisfied, it might even choose a forest clearing. Overgrown moor grass is too thick for it, so it has to move off into different meadows. Fertilized sweet-grass meadows crawling with prey for its brood are perfectly to its taste. If the bird is not "mowed away," it wouldn't have to leave the meadow until wind and rain flattened the grass. In France, the Corncrake owes its name, *râle des genets*, "broom rail," to the fact that the mower forces it to relocate to fallow fields overrun by broom.

A survey in Germany estimated eight hundred to three thousand calling male Corncrakes. It's not known how many pairs in the twenty breeding areas left in the country will even take steps to breed. In 1996, European Union laws protecting threatened birds prevented construction of a satellite village near Hamburg not far from a fen with fifteen Corncrake males; the same laws blocked the construction of a highway in the Enns River valley in Austria. EU laws trump national laws; however, we haven't heard the last word about the Enns valley. If there's any justice, that word should be *crex crex!*

The Royal Society for the Protection of Birds estimated that approximately twelve to fourteen thousand Corncrakes fall prey to hunters annually during their fall migration through Egypt.[26] And the bird is shot illegally during the hunting season in several Eastern European countries. There's no trick to shooting an awkward flier flushed by a dog. The bird's delicate flesh has always been prized after it has fattened and plumped itself up for migration. Nevertheless, *Crex crex* is nowhere better off than in Eastern Europe. If the species is considered "only" to be endangered worldwide, it's largely because of the economic crisis of the mid- to late 1990s in Russia. The country produced 220,000 tons of synthetic fertilizer in 1990; by 1995 there was only enough material to make 35,000 tons. Mower fuel is expensive, too, and so mowing in Russia is often delayed, if it takes place at all. Once the situation in the former Soviet Union improves, the Corncrake will be in for it there, too. If this miserable situation continues, the bird's breeding area will revert to woodland. In the meantime, Russian crakes, both "kings" and "queens," still rejoice, largely undetected, in their successive polygamy and wealth of whispering silky-velvety chicks, and bring the land whole armies of rasping bug exterminators.

CHAPTER 7

The Uncanny Goatsucker,
or the Willful Nightjar

A s bizarre as the bird itself is the ancient legend woven around its name. Pliny says: "THE CAPRIMULGI (so called of milking goats) are like the bigger kind of Owsels [i.e., blackbirds]. They bee night-theeves; for all the day long they see not. Their manner is to come into the sheepeheards coats and goat-pens, and to the goats udders presently they goe, and suck the milke at their teats. And looke what udder is so milked, it giveth no more milke, but misliketh and falleth away afterwards, and the goats become blind withall."[1] Silent as owls, the Nightjars, or "night shadows,"[2] glide back to their enchanted twilight realm after slaying udders, without giving the blinded goats a second thought. No wonder Aristotle, Pliny's source for the fable, emphasized the bird's uncaring nature.[3]

In all countries and at all times, Nightjars have stimulated our sense of fairy tale. Whoever lives in forest solitude, sleeps the day away, and is up to tricks at twilight or on a moonlit night must be in league with dark powers. The majority of fifty-five German names for Goatsuckers bear witness to this. Wherever cows in the pasture used to be driven by

Conrad Gesner, *Goatsucker* (*Caprimulgus europaeus*), 1617

children to their evening stalls, cattle and their keepers would call the swirling insect hunters "cow suckers" or "child milkers." A mysterious relationship between the bird and another dusk-flying, hunting, fluttering beastie—the bat—was compelling to other name-givers. They called the Goatsucker "king of the bats" because of its size and perhaps because of its superior silent flight as well.

A variant in the Nightjar's rich repertory of sounds is a long, drawn-out, plaintive *coeee-ek*, which the bird emits while searching for its scattered young. This call earned it the names "lament[er]" and "bird of the dead." The Goatsucker's purring sound, its constant mewing or squeaking, which were to our ancestors' ears like the whirr of a spinning wheel, sounded creepy in the dark woods. Was it a "witch" spinning flax on a lazy spinner? Did the "leader of the witches" show his flock how to produce an *oerrrr* sound when inhaling and an *errrr* sound when exhaling to imitate a continuous spinning sound? Today, the spinning Goatsucker triggers completely different associations. The sound reminds the Swiss ornithologist Urs Glutz von Blotzheim of "a small motorbike going by in the distance."[4]

Many ornithologists were uncomfortable with the superstition about Goatsuckers, which they could only comprehend as a denigration of the harmless "mosquito stabber." They would rather have had *Caprimulgus europaeus* still just called the Night Swallow, as was the long-established ornithological practice. Although the family of the *Caprimulgidae* is not related to swallows, most of the species in the large goatsucker family have swallow names in German. But *Caprimulgus* does mean, simply, "goatsucker," and so the name from classical antiquity won out. The French, like all Europeans, are familiar with the *tête-chèvre*, but give the lovely name *engoulevent* pride of place, which might be translated as "wind swallower."[5] In England, the Goatsucker's ornithological name is Nightjar, from *night churr* (the sound it makes), whereas Danish ornithologists prefer to call their *Gedemalker* the "*Nahtravn*," "night raven."

Why the "wind swallower" is also called "*Crapaud Volant*," "flying toad," in France, "goat milker" and "*Vliegende Pad*" in Holland, and why the German Goatsucker is also called "flying toad," "frogmouth," "night-clap[per]," and "flicker-off," is understandable when you inspect this idiosyncratic bird up close.

How magnificent is the Nightjar's large, curved eye, a dark, convex mirror amid soft feathering! Caught in the beam of a flashlight at night, its eye will glow red just before the bird shuts it reflexively, creating the illusion that it is only a dead piece of lichen-draped bark. This optical illusion is created by the bird's plumage pattern, which blends in perfectly with the surface it's on; if this happens to be a branch, the Nightjar will press itself down on it horizontally. Uncommonly thin-skinned and defenseless by nature, it relies on invisibility. It doesn't have to hide. One of the greatest trompe l'oeil artists in the bird world, it has mastered the trick of magically spiriting itself off stage. "So please do show us your Goatsucker," many of Oskar Heinroth's guests would ask. "It's right in front of one of you!"[6] was the astounding answer. The only signs of life that could betray the bird magically transformed into bark are its shining chestnut-brown eyes with their blue-black pupils, which is why this "daytime sleeper" seemingly keeps its eyes calmly closed—in fact, if you

Oskar Heinroth, *Goatsucker No. 1: 1. at 8 hrs.; 2. at 3 days; 3. at 6 days; 4. at 10 days; 5. at 13 days; 6. at 15 days, somewhat fledged; 7. at 18 days; 8. at 21 days; 9. at 24 days; vol. 1, 1926–33, photographs*

Oskar Heinroth, *Goatsucker No. 2: 1. at 30 days;*
2. at 33 days; 3. at 37 days, threatening posture;
4. 37 days, sunning; vol. 1, 1926–33, photographs

look closely, a fine crack, which Heinroth calls "biscuit-shaped," is always open.[7] That's all it needs for the clear-sightedness it doesn't lose even while asleep. Unlike an owl, it cannot turn its head completely around to the back to look behind it. Keeping stock-still is the strategy followed by this master of mimicry. It can stay snuggled down on a branch, a tree trunk, a root, or a forest carpet of needles and trust its eyes. Its eyeballs' legendary mobility extends its already wide field of vision up to its back. The narrow gap between its almost closed lids still permits the sleeper to perceive what's going on behind its back. Heinroth tested this out: as soon as he, silently and surreptitiously, held out a mealworm in a pair of tweezers behind one of his tame goatsuckers' backs, the bird whirled around and snapped up the prey thrown to it in the air.[8] To be able to see as if it had a third eye at the back of his head, the goatsucker merely had to turn *both* eyes backwards at the same time: "That way, the bird's facial expression was really terrifying; you're completely confused about where its head is and overlook its tiny beak more than usual."[9] Seen from behind, with its flat head, its eye slit, and the angled line of its back, it almost looks like a feathered frog, and that's not the only froggy thing about it.

The Nightjar's delicate mandible joint is highly maneuverable, just like its eye. The latter enlarges its angle of vision, the former its catching mechanism. A giant mouth hides behind that tiny beak. The deeply slit mouth opening in the bird's broad head runs back to behind its eyes. If "bigmouth" opens its beak, the two pairs of joints of the ramus of the mandible, which are divided into rotating and bendable segments, are pushed outward and enlarge the framework of the mouth's scoop-net;[10] once they are in there, the only way out of the soft pouch of the throat's skin for the moths, beetles, crickets, dragonflies, flies, and mosquitoes that have been swallowed alive is into their predator's stomach. Just in case some prey goes sailing past its beak, the Nightjar possesses an additional trapping tool, a sort of rake in the form of seven or eight stiff bristles on each side of the upper mandible; with this rake the Nightjar

Goatsucker, Dictionnaire des Sciences Naturelles, *2a. upper mandible, from above; 2b. claw showing small comb; 1816–30*

can sweep into its maw whatever the scoop—its beak tapers down to a narrow tip—doesn't catch.[11] When resting, the bird lays the slightly curved bristles together somewhat so that they stick out horizontally like a moustache on the upper lip because the points of the bristles lightly touch and point forward. If its beak pops open, the bristly traps snap apart and spread their tips. A toothed claw on its extra-long, outward-turned middle toe permits the Nightjar to comb the butterfly scales from its moustache after dining.[12] The tiny eyelashes on the edge of the upper eyelid don't need a special little comb.

Maybe it's been a long time since anybody has believed in goatsucking birds, but even today you can hear about the insect-hunting Goatsucker flying through the night with its beak permanently open like a winged scoop net. This notion originated with the bird's intimidation display. Only in the daytime and in the greatest of dangers will the Nightjar rely

Goatsucker, in Alfred Edmund Brehm and Emil Adolf
Roßmäßler, *Die Thiere des Waldes,* 1864

on the terrifying impact of its giant red maw. Its first reflexive reaction
when threatened is to freeze. Feathers lying flat, eyelids closed but for a
slit, it draws its head forward in such a way that, together with its back
and tail, it forms a single plane while pressing itself flat on its perch by
tensing all its muscles to look like a piece of wood with curiously soft
bark. If an enemy comes too close and the bird can no longer sustain this
illusion, it will open its large eyes, slowly raise its head, fully extend its
neck, spread its wings a little, and show its hellish, gaping gullet while
hissing and spitting. Neither friend nor foe will fail to be scared by this
display. Even an ornithologist expecting the threat is startled because it
is "extraordinarily reminiscent of the behavior of a disturbed cobra."[13]
As its gorge takes a few seconds to close, the bird's head is drawn back
down just as slowly as it was lifted. If need be, the spectacle of the hissing,
feathered cobra is repeated several times.

Among the Goatsucker's numerous enemies is the European adder
(*Vipera berus*). Only a serpent's wile can cope with a serpent. But it's not

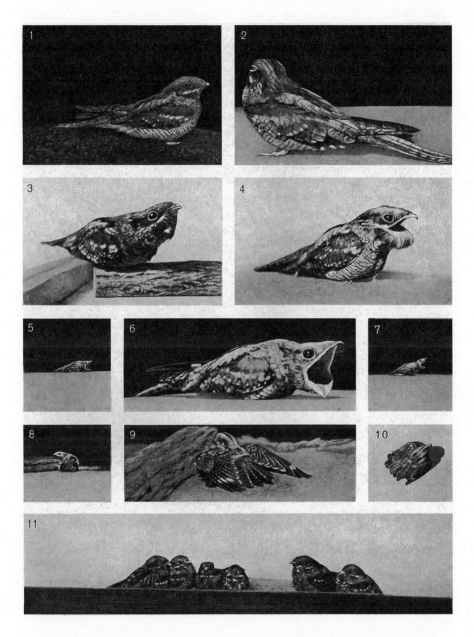

Oskar Heinroth, Goatsucker No. 3: 1. to 10. adult male; 1. roosting by day;
2. looking backwards; 3. and 4. initial, then, 5. to 7. full defensive posture; 8. about
to fly off; 9 and 10. sunning; 11. pair with young; vol. 1, 1926–33, photographs

recommended for a bird to pretend to be a snake in front of one, and the Goatsucker wouldn't dream of letting itself be hypnotized by a snake, like the proverbial rabbit would. In a nice role reversal, the bird entrances the aggressive adder. Fixing the enemy with an unmoving eye, it gets up on its short legs, deliberately raises and lowers its large wings, and slowly sways its body back and forth so that its fanned-out primaries rock in a cradling rhythm. Most adders pull back in a daze before this "scaresnake," and almost all of them lose their will to bite because of it.

No matter whether the Goatsucker is dreaming the day away, as Brehm states, or putting the fear of God into its predators, the moment the sun disappears below the horizon the bird stands up on its roost, jerks its head forward and backward, swings back and forth two or three times by shifting its weight from one leg to the other, and gets to work on its evening preening with its little beak: first comes the underside of its extended left wing, then its right; after a while, it tries out some takeoffs, landings, and pitter-pattering around, after which it sits down and once again savors the twilight. Then it takes off, "climbing high up in magnificent flight, as airy as it is graceful, and soars rapidly, now falcon-like, now swallow-like, claps its wings like a male pigeon, summons its mate with a gentle *he-eet he-eet*, flies playfully all around her with the most delightful twists and turns, and then descends on to a salient branch to sing or, more properly, to spin."[14] A bare, arm-thick branch of an old pine projecting horizontally from the trunk underneath a crown of needles—this is the preferred singing perch of the "night shadow," which sits parallel to the branch. It only goes higher up when it purrs; for a resting or hunting post, it prefers a branch only one to nine feet above the ground.

Even more preferable is sitting on a tree stump or on the ground and angling its head upward to scout for prey. Since it's best to see insects in dark woods against a somewhat brighter sky, the Nightjar always flies at them from below or from the side. Should the warm night air bestow it richly with prey, it will hunt in flight like a swallow: now hovering, pivoting on outspread wings, now shooting like an arrow "with drawn-in or powerfully beating wings."[15]

Bare spots and swaths cut in the woods, clearings, and paths are the Goatsucker's hunting and breeding territories. The site where the female lays her eggs, which the monogamous pair will incubate, must be sunny and the ground warm and dry. High grass and shrubs rising over heather and blueberries would only be a bother. The urge to build their nest in a concealed place is alien to them. They are concealment itself and manage without nests. Sandy and barren is the "night shadow's" world; the ideal environment for its magic arts is a pine heath forest.

The end of April or the beginning of May sees migrating goatsuckers return to their breeding areas in Central Europe from their African wintering grounds, at night and usually singly. The male occupies his territory with the occasional purr. The females arrive a few days later. A pair claims at least thirteen acres. The Goatsucker defends his territory against rivals until the beginning of July; after that time, other males might even take part in raising his young. In any case, he puts up with no other Goatsucker near him in May and June. He will pursue an interloper at high speed, calling and whipping his wings through the air with a loud, whiplike crack. At times, the male tries to fly right at the intruder, then glides by him to show him the shining insignia of his gender in the darkness: the white spots decorating the tip of his outer tail feathers and the three spots on the outer vanes of his primaries. The chase and bluster rarely fails to have an effect. If his rival takes off, the proud possessor of the territory once again resumes his parallel perch on one of his singing posts, where he can calmly purr to himself, unless he feels the urge to flick his fanned tail silently from side to side.[16]

Goatsuckers have been seen ruffling each other's plumage. If two fighters pull at each other's feathers with their small beaks—the fragile quills sit very loosely in their skin—they will lose some feathers, but won't sustain serious injury.

Pair of Goatsuckers, in Alfred Edmund Brehm,
Das Leben der Vögel, 1861

It takes up to three weeks after the female arrives for a Goatsucker's courtship display to begin. Purring he courts, and purring he copulates. The unending stanzas of his spinning song, the up-and-down swell of his *errr-oerrrrr-errrr-oerrrrr,* which he can keep up for nine minutes or more by vibrating his lower mandible, now no longer end in a breathy *kworrekworrekworre.* In love as he now is, his stanzas end with exultation that sounds like *djill… djilll… djillll* or *djeeyurrr… djyurr-djyurr… rrrr.*[17] This is sounded as soon as the desired Nightjar appears. If she comes to him in flight, he throws himself into the air in exultation, slaps his wings together with a far-sounding clap over his back, and entices the female to follow him with a *rueet* call.[18] The female, as if drawn by a long cord, likewise calls and claps along behind him. Before the pair can find themselves back on the ground, the male, floating down like a butterfly, spreads and raises his wings, fans his tail out wide, and presents to his mate out from his monochrome bark clothing, the shiny emblems of his masculinity, which are out of sight when he is perched. The muffled purring that an observing ornithologist heard afterward coming from the bird's landing site he could no longer see sounded to his ears as if copulation were taking place under water.[19]

It must have been a bright, moonlit night when another ornithologist succeeded in looking in on a Goatsucker wedding in the woods. The two birds sat on the ground facing each other: "Her body began to wobble back and forth, and his did as well. Then the female lay motionless, and the male changed his motion to an erect, strutting movement, jerking his tail and his whole back end up and down and partially opening his wings." His increasingly vigorous display behavior culminated in the sudden erecting and spreading of his tail with its eye-catching white spots: "All of a sudden the male raised himself up and with one quick wing beat glided onto the female's back."[20] With his trembling wings held high, and dropping his head with its vibrating lower mandible, the spinner copulated with his Nightjar.

"Love exerts its magic power even on the apathetic-looking Nightjar," Brehm comments on the Goatsucker's courting practices. He was charmed above all by the "night shadow's fiery" aerobatics during the "time of love."[21] We can thank Heinroth for a detailed description of *Caprimulgus europaeus*'s love and family life. He reared two Goatsuckers a year apart, and they mated and raised four young in his home. He called the adults Nora and Kuno. Both birds would lie like pieces of bark on their favorite roost, a leopard skin gracing a divan under a fan palm in a dining room corner. This little spot was Kuno's cozy roost as well as his singing perch. He presented Nora—she pitter-pattered or flew after him like on a drawstring—with nesting sites on a peccary skin on the

John Gerrard Keulemans, *Pair of Goatsuckers with Young*, 1897–1905

living room floor by propping himself up on bended wings and scraping the animal skin with his feet. It can take some time to scrape on a "forest floor" with tightly packed "pine needles." Nora had to frequently push her mate aside. Why all this scraping? I'm already sitting here—her way of letting him know that she seconded his choice was to settle down on the spot he had scraped.

Nesting in the living or dining room would have been dangerous for the birds and their offspring. Goatsuckers can make themselves invisible even in a furnished interior. They never sit against a light background, and when they stretch out to their full length, crouch down, and flatten their feathers because they're frightened, they look, if they happen to be on a little dark smoking table, like a supersized cigar.[22] The friendly giants Magdalena and Oskar often had trouble spotting their two

Goatsuckers and had to be constantly on the lookout so they wouldn't accidentally step on them. To protect the future babies, they put the peccary forest floor in front of an armoire in what was now called the bird room; the birds immediately took to it as their nest site. From now on, the living and dining rooms were off limits for the birds. Gone were the lovely days of unremitting spinning on the palm-sheltered leopard skin, of soaring through the suite of rooms, of posing like cigars on the smoking table. Now the job was to lay eggs, incubate them for sixteen days, and, once the chicks hatched, have two downy puffballs hanging from their beaks.

Kuno, lusting for copulation, would still purr when Nora was already sitting on her first egg. After unsuccessful attempts to approach her, he would sit silently on his corner of the armoire with his eyes glued to his mate on the peccary skin. The second egg came two days later. While Nora would leave the clutch for fifteen minutes or so, morning, noon, or night, to fortify herself on mealworms, cockroaches, and scraps of meat, or to stretch her legs or relieve herself under some cupboards in the farthest corner of the room—keeping the nest clean—Kuno would work on incubating—because you've got to learn how to do it if you're a male Goatsucker. How quickly an egg can roll away on a flat surface, and if you painstakingly roll it back with your beak, the other one might well have rolled away in the interim! It took some time for Kuno to figure out how Nora eased herself onto the clutch by supporting herself on bended wings, while shoving her feet under the eggs so that she rested on the top side of her toes.

The birds were not annoyed by their shutterbug foster father. The incubating female proved to be particularly reasonable: "The peccary skin was directly in front of an armoire, and when unavoidably opened it often forced the good mother to duck to let the door pass over her head!"[23]

On June 18, Heinroth noticed a tiny break in the first egg and heard a thin, peeping sound. He gives a very graphic account of how the little

Oskar Heinroth, *Goatsuckers: 1. at 7 days;*
2. adult male; vol. 1, 1926–33, colored photographs

Goatsucker broke through the shell: "It burst [the shell] the same way that probably all birds do by using its egg tooth to break through 'the Tropic of Cancer' from the inside so that it could then lift off the 'northern temperate and polar zones.'"[24] The first break on the egg globe's Tropic of Cancer was visible around noon, but the little polar cap wasn't lying beside the incubating mother until early morning two days later, "and a quarter of an hour later saw the rest of the empty shell beside the nest site."[25] The little downy puffball weighed about one-sixth of an ounce and its contours made it hard to recognize as a bird. When Heinroth picked it up fifteen minutes after it hatched, the "little citizen of the world" was already dry and merrily crawling around in his hand with its eyes open: "The beastie is but a few hours old when it responds to a soft *curr, curr, curr* of the bird sitting on it by quickly emerging from under the adult's breast feathers; then it turns around, straightens up on the adult's breast until, finally, its beak grabs its father's or mother's beak. Once it has seized the beak, it holds on rather tightly; in fact, it can even happen that, if the adult raises its head, the little one is left hanging on to it for a little while because it won't let go of the parental beak."[26]

The adult birds do not cut up or predigest the young birds' insect fare. It is mixed with saliva, pressed into long balls, and pushed down into the little "beaksuckers'" gorges. As with most species, the young accept their food with wings slightly spread, and their whole body shivers. After feeding they pitter-patter quickly backwards, deposit their feces, and trip back just as rapidly to the incubating bird. Since they always trip along backwards the same distance, as if hitting an invisible obstacle, and in all directions, a wreath of little piles of white excrement accumulates over time around their nest; and because the number of little steps, constant for a while, increases as the chicks grow bigger, the rings around the smallest wreath grow along with them unless the family is disturbed and moves to a new nest site that will have a larger ring grow around it. As soon as the offspring can catch their own insects, they evacuate as regularly as the adults do.[27] Fairy rings (*Hexenringe*, or "witches' rings") is the name given to the circles that many cap mushrooms create around a common mycelium that germinates underground, radiating in all directions. The "leader of the witches," or the "witch," as Goatsuckers are sometimes called colloquially in Germany, sits in the center of the rings of white feces. Why their offspring deposit their excrement in such a magical fashion is one of ornithology's unsolved riddles.

The second little Nightjar freed itself from the shell on the peccary skin twelve hours after the first hatchling. This chick also had a coat of down that was dense underneath for the cold forest floor and sparse on its back for the mother's warm brood patch. The adults' deep grunting sound when brooding is like a *wuff-wuff-wuff*; the chicks' whispered begging song, a *brrree... brrree*, a development from their original thin peeps. If the parents don't come up with some food, the new additions to the family will give them sharp tugs on their beak bristles.

Heinroth couldn't yet know in 1908 that Nightjars, in the short time span they're on territory, always manage to pull off a second clutch that overlaps with the first brood. He saw with dismay that Kuno was once again spinning and flapping his wings after the squatters had hatched and was wooing the female; and soon enough, she requited his desire so that the male's "weird murmuring"[28] during copulation was heard once more. The ornithologist feared that the love-crazed pair might neglect their progeny or even abandon them, but it turned out that his fears were groundless. A photograph shows how the birds' love for each other included love for their young. Beak to beak and breast to breast, they sit in beautiful symmetry opposite each other and each one has a seven-day-old chick under it.

Ere long, two more eggs were lying on the old nest site on the peccary skin: "The young were now thirteen days old and had always been tended by their parents, and they would of course try to crawl under their mother that was already incubating again, but she didn't seem to take kindly to this intrusiveness."[29] Then Heinroth would protect the breeding mother from her offspring by day with a wire bell jar for keeping off flies, so that the chicks, at roosting time, had to be content with the shelter of the paternal brood patch and would usually tap Daddy's beak to still their hunger.

In the woods, the second brood begins to hatch on the very day that the chicks from the first hatch enjoy their independence. At twenty-three days old, the two offspring of the hand-reared "parlor goatsuckers"— which, at best, could catch a fly—were fledged, completely feathered, and still fond of Kuno's beak when their peeping siblings surfaced. Freed from having to get food, the parents were able to stuff all four beaks together.

Heinroth writes:

It was of course a great delight for us to observe goatsucker family life in the twilight hours from such close range. The littlest young ones scooted around the floor sort of like chicken chicks on half-feathered feet, energetically flapping and giving each other their regards. The parents were busily sashaying around . . . and were often urgently pursued everywhere by the young from their first brood when they took wing. The animals were never bothered by our presence and often sat on our heads and shoulders or hovered and shook while begging at our hands.[30]

Its knack for standing and hovering almost vertically in the air for a few seconds lets the Goatsucker look closely at strange objects in the woods. Whatever might make the bird freeze by day can be examined in good time at dusk, by moonlight, or in the gray of dawn from its viewpoint in the air: a looming deer, a sniffing dog, a bicycle leaning against a tree along with the ornithologist owning it; and because it is disconcerting to the Goatsucker that this man suddenly throws his hat in the air and then his walking stick, he will try to catch these things as he's shaking and hovering.[31]

The adults react completely differently when a person approaches their nest. They will fake a wing injury every time and in every which way, but in the nighttime the Goatsucker that was just now creeping along the ground with twisted wings will all of a sudden fly up and flutter around the interloper's head within reach, while emitting its warning cries, *gritt-gritt-gritt* or *dack-dack-dack*.[32] That's why one German dialect name for the Nightjar is *Wegflackern*, roughly, "flicker off," "flutter away," because *flackern* originally meant "to flutter."

Whether or not Goatsucker offspring meet one another in the woods with the same politeness as in Heinroth's bird room is not recorded:

> Until the young birds are independent, they have a very curious way of greeting each other, which is exhibited less often with their parents. If, for example, the young in the room are at some distance apart and then meet again, they run to each other with wide-open, upheld wings while uttering strange murmuring sounds. Their body and their head's longitudinal axis stay precisely horizontal, and the animals make for a most unlikely picture in this pose. We never observed this ceremony later on, as it is not the custom between mated birds.[33]

Heinroth observed feathered chicks at fourteen days making fluttery, hand-high jumps; a week later, they were already fully feathered and flying up onto the armoire. The little ones imitate a hissing cobra, in the forest as in the parlor, even before they've fledged, whenever they feel they are under threat.

Heinroth felt that the Nightjar's mimicry program included "transitional movements."[34] How can you stay invisible in the light of day if you're faking lifelessness when incubating and nevertheless have to move away because you're frightened off? You can't simply fly or run away, so you act as if you're blown by the wind. The Nightjar sits immobile on the bare ground. Its head buried in its feathers, it begins quite subtly to oscillate its upper body for a brief while and then stops, gradually rocking more and more, as if in the wind, which apparently stops as imperceptibly as it came up, since the Nightjar once again has stopped moving. Then when the Nightjar, rocking harder, moves toward its nest of eggs, it looks like a mere piece of bark or a little pile of leaves moved around by the wind. Very gradually, the same way it started, the rocking dies out once the wandering shred of bark has finally covered its eggs again.

Reiner Schegel has his doubts about Heinroth's mimicry interpretation of this ten-minute wobbling game, which he himself was able to observe a number of times in the pine heath forest of Upper Lusatia (Oberlausitz). This Nightjar researcher asks, "But where's the protection if an enemy were actually to observe the process?" He believes that with its minutes-long rocking, the bird "draws more attention to itself than if it would sit still on the nest at once."[35] That may be the case if foxes and martens are involved, but they prefer to hunt at night or at dusk anyway. In the daytime, the Nightjar's chief predators are the Sparrowhawk (*Accipiter nisus*), the Goshawk (*Accipiter gentilis*), and the Peregrine Falcon (*Falco peregrinus*). If these birds are watching for a sign of life in their prey, their sharp eye could totally overlook the bird in the piece of bark blowing in the wind. But if the enemy is a human, the Nightjar cannot just "rock him away," especially if it is an ornithologist, of all people.

In spite of having two broods and non-breeding males helping feed their young, rarely do more than two Nightjars fledge from a season's four eggs. Not only martens, badgers, foxes, hedgehogs, and wild boars relish birds' eggs. The tiny Eurasian or Common Shrew (*Sorex araneus*) appears not to eschew fine food either.[36] With its sharp little teeth, it can gnaw a hole in the shell as wide as a finger and then slip into the pristine enclosure and lick it clean. Mother Nightjar has no idea that a furry animal is living in an egg and after him, nothing.

Nightjars fly at low levels to catch soft-skinned flying things like mosquitoes for their newly hatched young. Their scoop nets catch larger or smaller insects depending on the weather and season. From rusty long-horn beetle to stag beetle, from cockchafer to burying beetle, from silver Y

Whip-poor-will.

John James Aububon, *Whip-poor-will*
(*Caprimulgus vociferous*), pl. LXXXII, 1827–38; the Common
Poorwill (*Phalaenoptilus nuttallii*) is somewhat smaller.

moth to large yellow underwing, from pine hawk-moth to goat moth, from nun moth to garden tiger moth—there's room for everybody in there. The bird's giant maw suggests insatiability, but that's a fallacy. Compared with other birds of its size, the Nightjar is rather frugal. Thirst is something it doesn't seem to know; at least our experience with captive birds has taught us that they maintain their metabolism without a direct supply of water. When Schlegel gave one of his tame goatsuckers some water, it shook it off. If he held the bird's beak shut to force it to swallow, "a sort of 'slight digestive trouble' regularly followed, manifested as frequent deposits of thin, watery, foul-smelling excrement. Otherwise the feces are firm and hardly smell."[37]

A bath in water, so pleasing to other birds, is resisted by the Goatsucker, which would rather take a shower. Heinroth's Goatsuckers would blissfully spread their wings and tails at the open, grated bird-room window.[38]

Most enjoyable for the Nightjar is a sunbath. It spreads its tail so wide that it forms a unique, almost square field of feathers together with its wings, which are spread forward under its head to trap the sun's rays. In this position, it is extremely easy to detect, so it chooses to keep its big eyes open, which can even stare into the noonday sun.[39]

If food is scarce during cold, rainy weather, the Goatsucker proves to be a hunger artist. One of Schlegel's well-nourished alumni went eight days without food and apparently fell into a sort of torpor. Its normal body temperature in a cool room would have been 100.4 degrees Fahrenheit. In his lethargic state, it sank to 57.2 degrees; its breathing motion was barely perceptible, and its bloodless mouth looked whitish-gray. The bird sat there as if dead, with completely closed eyes, fluffed-up feathers, and slightly spread wings; it did not react to any touch at all. With the approaching darkness, its temperature spontaneously rose again and reached room temperature after a good hour. The Goatsucker was compensated with much food after surviving this risky experiment, as were his captive species mates that were put to the hunger test later.[40] Because their torpor thawed in the evening, a metabolic slowdown in nature cannot last longer than one day.

What the European Nightjar cannot do, an American relative—*Phalaenoptilus nuttallii*, the *Winternachtschwalbe*, "winter(ing) nightjar," called the Common Poorwill after its call within its home range[41]—can. A bird of this species once started to spend three winter months in southeastern California squeezed into a rocky crevice in a Chuckwalla Mountain canyon; it was in a deathlike torpor when three American ornithologists discovered it in December 1946. If the bird had not been there, we might still believe today that hibernating birds were only found in old wives' tales. For four winters in a row, Edmund C. Jaeger observed the bird, which stayed loyal to its crevice in the rock. Its body temperature was 65.3 degrees Fahrenheit; a mirror held up to its beak didn't cloud up; a stethoscope didn't pick up a heartbeat, and a beam of light directed at its eye didn't provoke a reaction. In spring, the Common Poorwill gradually thawed out and woke up and shook itself violently as if waking from being drugged. In a cold-temperature experiment, Common Poorwills were able to lower their temperature to 41 degrees. *Hölchoko*, "the sleeping one," the Hopi people have called it since time immemorial.

The Nightjar plays an important part in the mythical universe of the North and South American Indians as a creature from the realm of the dead. Regarded by many as a friendly messenger from the deceased, by others as the vampire soul of the dead, as the servant or relative of the water genie, as one of the four divinities surrounding Moon, the goddess of vegetation, or as the lady of destructive fire, of fever, cramps, thundershowers, and honey—the natives of Guyana looked upon her as being four-eyed, whereas in southern Brazil, Paraguay, and northwestern Argentina, people believed a sleeping bird left its eyelids open a crack to not let the sun out of its sight because, before having been transformed into a Nightjar, it had been a moon being that had lost its lover, the Sun. In the Amazon region, it was the custom to sweep the ground under young girls' hammocks at the onset of their puberty with Goatsucker feathers

Anita Albus, *Goatsucker before a Burning Forest,*
watercolor and body color on tinted paper

to protect their virtue—the hammocks were garnished with a Nightjar skin, or the girl had to sit on a feathered skin for three days. The Brazilian Indians often compared the Goatsucker's huge mouth to a vulva, one of many variants showing the bird as representing oral greediness. Whether Mrs. Goatsucker, as Moon's wife, serves him only green squash because she gobbles up the ripe ones; or, as the joint spouse of Sun and Moon, she sparks quarrels and jealousy between the heavenly bodies because the bird flees Moon's chilly touch and would rather enjoy Sun's warm embraces— ultimately she is always abandonment personified and has been mourning her fate ever since on moonlit nights. Mr. Goatsucker shares the same fate since he sought in vain to seduce Lady Moon, who was completely devoted to Lord Sun. And how it happened that potter's clay came to the earth when a female Goatsucker fell, abandoned, with the clay inside her, or not, while her spouse-planet rose to the sky—all these myths can be read about in Claude Lévi-Strauss's *La Potière jalouse.*[42]

Even now, Goatsuckers come from tropical Africa to their Central European breeding grounds; even now, they can be heard purring wherever knotted pine, elderberry, and heather flourish on sandy soils. But their spinning frequently lasts so long that ornithologists doubt whether they will have a family. Many of their survival strategies have, in the modern world, become their doom. Streetlights and the warmth of asphalt streets at night attract insects and their hunters along with them. Goatsuckers' reflex reaction to rest when a strong light hits them turns them into traffic victims, particularly during spring and fall migration. Even steps taken to protect nature are destroying goatsucker territories. A Nightjar cannot breed where a Plenter forest (*Plenterwald*), a continuous cover forest with a tight, leafy umbrella, replaces cut-over land. On the *Red List of Threatened Vertebrates in Germany, Caprimulgus europaeus* was recorded in 1994 as "endangered." In the meantime, its status in the whole of Central Europe has worsened.[43] As a forest dweller, it was already lost to us long ago. Only to the Indians has it been granted to unfold in their myths all the things the "night swallow" can do.

CHAPTER 8

The Beautiful Barn Owl

T HERE IS no owl that does not have a veil. All the species have one, be it the Eagle Owl or the Screech Owl, the Scops Owl or the Snowy Owl, the Pygmy Owl or the Dusky Eagle Owl. The German term, *Schleier* (veil), is misleading in so far as the "veil" forms the face of the Barn Owl (*Tyto alba*) rather than covering it. *Disque facial* is the French term for it, as is "facial disk" in English. The owl's facial disk resembles a veil in two ways. First, the small feathers of the front veil— white in the Barn Owl—are so fluffy that the stiff barbs, which branch out separately from the shaft two one-hundredths of an inch apart, are layered on top of one another to form a veil-like grillwork. Second, the disk's shape as a whole "veils" the avian form of the skull and lends the owl a seemingly human face because its eyes face front. The most perfect "veiling" of their forward-pointing heads is found in the thirteen species of the Barn Owl family, from the Madagascar Red Owl to the Moluccan Hawk Owl, from the African Grass Owl to the "heart owl," a colloquial German nickname for *Tyto alba*. A closed, heart-shaped veil of silky, shiny, powdery-white feathers frames its blackish-brown eyes and narrow bill, which imparts a certain human quality to its face. The front, or inner, white veil sits beneath the golden-ochre pointed tips of the outer veil behind it, curving from the middle of the forehead down on either

John James Audubon, *Barn Owls (Tyto alba)*, pl. CLXXI, 1827–38

side past the large ear openings to the lower jaw. These extremely close-packed and narrow little feathers, unlike the latticed feathers in the front disk, are arranged in eight rows one behind the other, framing the heart-shaped face like a kind of lace ruff. The bill's base is concealed by feathers, and the part of the veil above it is folded over like a bag to suggest a nose, the ridge of which looks like an angled, two-sided border to the left and right halves of the disk; some rusty brown feathers in a row running from the inner corner of the eye to the bill hint at a line on the side of the nose. Anyone who regards animals as incomplete preliminary stages on the path to the perfect human form, as did the natural philosopher Lorenz Oken, would have to believe that the Barn Owl is "incontrovertibly" the most beautiful owl of all.[1]

No human being can match a Barn Owl's clear, long-range night vision, and no human ear is able to hear what the owl's ear can. The sound of bells doesn't faze it if its nest is in a church tower. It's got nothing to do with the bird, and the thundering noise is much too conspicuous for its finely tuned hearing. Its ear is pitched to high broadband frequencies, to the twittering chatter of shrews staying in vocal contact with one another; to the warning whistle of true mice; to its rustling in dry leaves or hay; to the sound of its chewing; and who knows if it can even hear the grass grow. It has the best ear of any bird in the world. If Barn Owls don't go out hunting during a heavy rain, it's not only because they're afraid of getting their feathers wet and finding it hard to fly but because they are frustrated on account of their hearing: the rustle of the rain and the splattering of raindrops on their feathers drown out mouse sounds so that the owls can't locate their prey. Any other time a mouse makes a telltale sound, it can hardly ever escape the owl no matter how black the night, because the owl is "all ears" thanks to its veil. The excellence of its veil equals its legendary sense of hearing, surpassing that of all other owls. The two halves of its heart-shaped face form two auricles, or external ears, when the bird projects them forward in a circle in order to listen; they become two parabolic reflectors that focus sound. For now,

Anita Albus, *Barn Owl Feathers from Front and Rear Facial Disks*
(actual size 1¹/₁₆ inches), watercolor and body color on paper, n.d.

the Barn Owl still has a secret about the thick feathers in its outer veil, which has barbs branching off laterally from flexible shafts, barbs with up to twelve tiny barbules every three-hundredths of an inch. The mystery is this: how come the outer veil feathers reflect sound and don't swallow it up?[2] Ornithologists also puzzled for a long time over a skinlike, feathered membrane in front of and beside the eye and in front of the ear opening; it is like a flap, which would seem to screen out noises coming at the owl from in front. A student at Cornell University experimenting with Barn Owls helped us understand how the flaps and all the parts of their complex auditory system function, including the apparent nose—the white, feathered wall that prevents sound waves from slipping over from one parabolic dish into the other.

Roger Payne used an infrared camera to demonstrate how Barn Owls can instantly catch a mouse rustling around in a room so completely dark that even the owls' telescopelike eyes are unable to see anything. A rustling noise would come from leaf litter. A slight draft from the owl's silent flight as it pounced on the mouse would be the signal for Payne to turn

Oskar Heinroth, *Barn Owls: 1. at 6 days;*
2. nestling at 35 days, in second downy plumage;
3. adult; vol. 2, 1926–33, colored photographs

on the light. He discovered that the owl had a mouse in its claws every single time. If he took away the leaves and tied a leaf or a crumpled piece of paper to the mouse's tail, the owl would hit the sound but miss the mouse. Neurobiologists expanded on his experiment and determined that Barn Owls have a neuronal system in the midbrain that depicts sound three-dimensionally in the form of a spatial map.[3]

The asymmetrical, laterally projecting ear flaps are needed for the owl to find its bearings by sound in the dark. They adjust the auricular opening so that the ear receives back the sound waves passing through the front of the veil—waves that are ten times stronger when they leave the focal point of the parabolic dish. The left flap projects more above

Oskar Heinroth, *Barn Owl No. 1: 1. at 18 hours, in first downy plumage; 2. at 27 days, in second downy plumage; 3. at 35 days; 4. at 42 days; 5. and 6. at 47 days, dropping feathers; 7. at 54 days, able to fly; 8. at 61 days; 9. and 10. at 79 days, fully grown; 9. active, ready to fly; 10. drawn up in alarm;* vol. 2, 1926–33, photographs

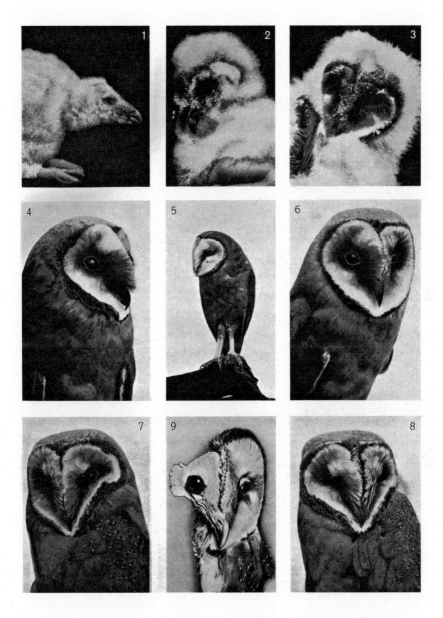

Oskar Heinroth, *Barn Owl No. 2: 1. at 16 days; 2. at 27 days;
3. at 42 days; 4. through 6. adult, active; 7. somewhat alarmed;
8. more alarmed; 9. dead Barn Owl with right side of head bare, showing large
front ear covering (tagus) behind the eye; vol. 2, 1926–33,* photographs

the skull than the right flap, and their edges are at about a fifteen-degree angle. This asymmetry produces a difference of a fraction of a second between the sound hitting one ear and then the other. The owl's brain can read out the vertical and horizontal distance to the source of the noise using the differential between the frequencies in the near ear and the back one, which is placed a minimal distance behind it. The bird's three-dimensional hearing gives it such an exact fix that the margin of error in both the horizontal and vertical planes is less than one degree. This does not preclude some prey in the wild escaping the owl at the very last second. Although a mouse cannot hear an owl's flight either, the draft from its more than three-foot wingspan is transmitted more clearly to that little animal than to the experimenter in a dark laboratory.

The owl's soft and silent flight comes from the pliancy of its feathers. The airflow dissolves into very fine turbulences because of the delicate velvet layer on the upper surface of the feathers' vanes—the fine barbs of which are connected by even finer barbicels—which terminate in long fringes. The serrated leading edge of the outer primary wing feathers contributes to this feature as well, meaning that the sound of the prey is not masked by any sound the owl might make.

An hour after sunset, the Barn Owl opens up its listening dish and goes off hunting. The graceful ease with which its broad, soft-feathered wings move as it slowly patrols low over open country endows its stalking flight with a butterflylike quality. Should it hear a suspicious noise, it can reverse direction with lightning-quick speed and instantly switch from soaring to rowing, from hovering to diving. If the weather is bad and it would rather hunt from a perch in a barn, it can easily take off vertically or backward among the angled beams.[4] It likes to listen for the sound of its prey by changing its listening post minute by minute—from a tree branch to a bale of hay, a fence post, gutter, windowsill, wall, or stone vase on a gatepost, where its silhouette looks entirely natural in the dark: a vase on a vase. When the sound of a mouse reaches its ear, the owl slowly flies in a flat arc, with deep, deliberate wing beats, in the direction of the sound, adjusting its flight to its prey's shifting location. Hovering in the same place and dangling its legs will retard its flight and help it get its bearings by ear. With its sound visor to the fore and legs pressed tightly against its tail feathers, it makes a sudden dive, puts on the brakes before impact—wings tilted sharply upward with alulae (small "thumb" wings) outspread—retracts its head, turns 180 degrees at the last second, and yanks its claws—now joined together like a pair of tongs—to align them with its facial disk and the intersection of its acoustic crosshairs. When it strikes, its eyes are closed. There is no escaping the embrace of its eight widespread toes and sharp claws. A well-placed bite on the nape of the neck, and the mouse is dead. The farsighted owl has to feel for the nape by gnawing, which is why it can easily afford to keep its eyes closed. The long, flexible, antennaelike hairs at the base of its beak—the vibrissae—are useful for registering things close up. The owl spreads its wings over its prey like a mantle—this is actually called mantling—and guards the prey from behind with its tail. No jealous competitor is going to snatch it away.

If a mouse shows its face, the owl's eye will guide the bird as it makes its catch. Moonlight is unnecessary; the owl can do very well in quite low light as it swoops down on its prey. It closes its eyes before impact, just as it does when hunting by sound. The owl's uncommonly large eyeballs are not spherical but are projected longitudinally and sit like a bell in the cup in the back of the eye. (In German, *eyeball* is "*Augapfel*," literally "eye-apple.") To stay with the fruit simile, we could call their eyeballs "eye-pears." They have a pupil that can dilate to cover the entire eye opening; behind it is a large, very convex lens. The eye is also marked by the wide-angled opening of the cornea and the cornea's large curved surface compared with the relatively small surface of the retina. The owl's eye is made for the night, so it can concentrate what light there is into a small but well-illuminated retinal image without any diffusion. The retina of

Tyto alba, the most nocturnal of owls, produces the brightest images. Its retinal cells have even fewer cones for perceiving color than do other owls' retinas, but the longer and slimmer rods are all the more densely packed for light–dark adaptation. Young owlets learn as they grow older to connect their visual perceptions with acoustic ones. That is how they develop a harmonious combination of neurons in their visual and auditory systems, a mutual coordination of the auditory and visual spatial maps represented in the tectum of the midbrain. In the above-mentioned experiment, the Barn Owls' vision was completely blocked; in normal life they would of course always use all their senses at once. Even if they cannot see or hear anything, for example, in a belfry at night, they can still fly, guided by their superb sense of local memory.[5]

The Barn Owl's eyes are relatively small for an owl. Its facial disk hides the eyes' forward placement in the skull, in which they are firmly anchored like a telescope. Its ability to rotate its cervical vertebrae exceptionally far makes up for the immobility of its eyeballs. It can move its facial disk to face backwards with lightning-quick speed, as can all owls, when, for example, it hears a noise coming from behind as it lies in wait outside a window, even if it's the soft sound of a scribbling pencil. The owl's wig of thick reddish-blond and gray feathers, decorated with streaks of small black-and-white eyelets, will then be facing forward in lieu of a face, which simultaneously conceals its long, thin neck.

To help them connect their visual and auditory perceptions, young Barn Owls practice getting their bearings by turning their head vertically ninety degrees so that the bill in their droll heart-shaped face is horizontal and one eye lies directly below the other. But it looks completely crazy when they keep turning their head vertically to reach 180 degrees, literally head over heels ("neck over head" in German). Their beak then points upward almost vertically over their eyes, as if a sloppy hangman had stuck their disconnected head back on their shoulders upside down.

Anita Albus, *Barn Owl "Head over Heels,"*
watercolor and body color on paper, n.d.

The size of a Barn Owl's clutch and the number of broods depend on the available mouse supply. The owls react to a rise in the mouse population with increased fertility. Barn Owls can lay up to sixteen eggs in one brood. Underneath their luxuriant plumage they are very skinny and suffer high mortality in the winter cold when mice make themselves scarce. They are unable to build up fat reserves. In years when mice are plentiful, the Barn Owl averts the demise of its species by producing numerous overlapping broods and spreading the breeding season over eight months. If there are no mice, there are no offspring either.

The female's plumage is a shade darker and more closely delineated than the male's. The male is more delicate, lighter, and smaller, as is common in the owl kingdom. This is why caution is the order of the day when he sets out to win a female. He must be on guard for quite some time because of her stronger beak, her bigger claws. His burry shriek while marking his territory and showing off his breeding territory to his *belle de nuit* reminds us more of a rattle's racket than of birdsong. If a rival has the audacity to encroach on his territory, perhaps to usurp his breeding ground, the owl's scream would make your blood run cold. If screaming and chasing aren't enough, the rival will be attacked in the air. Claw-to-claw duels can end fatally, but usually the weaker bird, or the smarter one, takes flight before any serious injury occurs.

Barn Owls are the only birds in the owl kingdom that copulate in the nest. A niche in a church belfry is their most coveted nesting site because the stone (or beech) marten (*Martes foina*), a dreaded egg slurper and chick devourer, cannot climb that high. Scraping the ground, hopping about, and flapping his wings, the male lays claim to his breeding site and uses his beak to break up the thick felt of pellets his forebears have left behind. A stock of mice will demonstrate to his future Mrs. Owl how provident a *père de famille* he is.

He must belt out his shivaree song from the tower window for a long time until, after many nights of waiting, the Owl of Owls finally appears. There she sits, on the ridgepole of the nave, acting as if she doesn't even see him. The very sight of her lends him wings, and he invites her into his love nest with a soft, throaty purr that swells up into a shrill buzz while he constantly goes in and out of a vent in the tower. But his unapproachable belle wouldn't dream of hopping onto the pellet carpet with a male owl at the drop of a hat. She first wants to have a look at how skillfully he flies, how well he can fight, how much of a hunter he is. The moment she takes off, he's already in the air, and she goes after him in a furious chase. After several laps around the church tower they land next to each other on the roof and purr a duet. Before the circling pursuit resumes, they strike each other on their facial disks with their beaks.

For nights on end the male tries in vain to entice the female into his tower nest. When he finally succeeds, it's because of his tireless

Anita Albus, *Barn Owl with Pellets,*
Alarmed by Day in Attic, oil on copper, n.d.

stockpiling of mice, a procedure that the female observes in flight. The shivaree invitation now sounds forth from the nesting niche. Soundless the female's flight, soft her landing, finely tuned the male's ear. The female has barely touched down when the male, answered at last, storms over to the window and welcomes with a purr his bellicose bride.

She drops by more often now to visit him in her future abode, but woe to him if he approaches too closely! His beak is no match for hers, and there isn't a visit from his "heart owl" that doesn't end in beak-fencing, unless he raises his wings in a conciliatory gesture the moment she unfurls her threatening facial disk, backs away, bowing and scraping like her most humble and obedient servant, and then beats it. He can count himself lucky if she doesn't go after him anymore. After some scratching and gnawing among the pellets she can rest assured—by using the sensitive soles of her toes and the vibrissae girding her beak—that he has not been gorging on the mice he has brought to the tower but has stored them up for her. This seems to put her in a more gracious frame of mind. If the male now appears in her close proximity, she does not immediately drive him off. The two birds stand facing each other, bow with wings slightly raised, and hum a duet while slowly prancing around in a circle. Since they can read each other's feelings from their facial disks, they stay turned toward each other. At the slightest dissonance the round dance of the purring duet turns into a violent beak-and-claw fight. Purring, chirping, shrieking, they scuffle, but no serious harm is done.

The stalwart Mr. Owl has given it his best shot. He has defended the safest breeding site in the whole village against all comers; sung and wooed long night after long night; laid up a store of a good dozen mice; demonstrated his aerial artistry and his fearlessness in battle with a stronger female—but his toughest task still stands before him: the Taming of the Bride. She evidently has a weakness for the dance he regularly performs on the nesting site. The moment she appears, he assumes a stiff pose, bows his head, and performs a slow tap dance among the pellets, purring softly at first, then louder and louder while stamping on the floor at brief intervals with a jerk of his long legs. Hissing, tongue-clicking, and beak-clattering heighten the infatuating effect of the purring and stamping music of this slow-motion flamenco.[6] Ornithologists call this the "nest-site display." But even after all this, the male has only half-won the heart of his *belle dame sans merci*. He does not become irresistible in her eyes until he combines his dance with a love gift: a newly killed mouse. The female must first of all bow and take the mouse off the floor before his feet. As soon as he is allowed to pass it from beak to beak while uttering a tender chatter, he has won the game.

If the stronger owl grows weak, the weaker one can finally be the stronger. Almost four weeks of brawling courtship have gone by—a long time in a Barn Owl's usually brief life. The harmony that now reigns could not be greater. The most sharp-eared and farsighted among nocturnal owls represents in its love life, too, an escalation in the bird kingdom. Its so-called "cloacal kiss" is not limited to a few seconds, as it is with other owls and birds. When the male starts up a tender chirruping, the female crouches down flat among the pellets, closes her eyes, and sends a buzzy cheeping as an invitation to mate. The male rushes over straight away with a staccato cackle and uplifted wings. He half-flies, half-mounts onto her proffered back at shoulder height. Once he has fluttered around to find his balance, he squats on his heels on her back and bites into her neck feathers, supporting himself with one or both wings, while the female angles her tail upward and to the side so that he can sink the base of his tail onto her cloaca, or vent. As he rhythmically presses his vent against hers, his staccato cackle and her buzzy cheeps increase in the same rhythm. Copulation can last up to sixty seconds and concludes with a powerful, orgasmic push, accompanied by the male's snoring crescendo; then he jumps off, emitting a coo as his excitement dies away, while the female waits for several seconds and then slowly sinks her tail. The birds thoroughly ruffle each other's feathers afterward, drawing every single latticed feather of their facial disks lovingly through their beaks. Every now and then they drop off for a while, breast to breast.

Now the female spends most of her time preparing a soft nesting hollow. She carefully chews some pellets while taking her time to scrape the nest into shape. Meanwhile the male keeps repeating his stomp dance for her and doesn't forget to serve her up a mouse afterward, from beak to beak, along with some chattering. Holding her love gift in her beak and humming away, the female offers her backside to her cackling spouse. After they copulate, the mouse, which has been swallowed lock, stock, and barrel, tastes oh, so good. Once in a while, the female will go out hunting herself. Two or three weeks go by like this, with the birds copulating only a few times at night. And when the female leaves the nest site just to remove the feces and when she, like a baby owlet, begs the male for food and makes a snoring sound, then "copulation thrives." Every feeding is followed by a mating; in between, the male goes hunting and does his dance. Now he hangs around her during the day as well, and she calls upon him to copulate at least once every hour, day and night. They ruffle each other's feathers for a long time afterward, cuddling up close. Dancing, feeding, copulating, tender preening, and sleeping disk to disk, interrupted only by the male's nightly hunting excursions—this fills their nights and days at the climax of their loving courtship.

About four weeks after their first mating, the female owl pushes her first egg into the soft, warm hollow among the pellets and covers it with the unfeathered brood patch that has formed under her belly. Spoiled by her spouse with mice and liberated from hunting, she has been transformed into a portly setting hen a good third heavier than her mate. If there were a huge surfeit of mice, she would get even plumper and under certain conditions would grow half again as heavy as the male. Sixteen eggs a little over half an ounce each would just about add up to her normal weight.

The Barn Owl's white eggs are more oblong than those of other owls, and their delicate shell has no luster. The owl lays them thirty to eighty hours apart. She is the sole incubator, whereas the male provides her with food, copulates with her just as often while she's on the eggs, and preens her feathers. Ornithologists still have not discovered how the owls know how many offspring they can manage. Whatever the male cackles into the female's ear during his staccato of copulations he chooses to keep to himself. The female probably knows from the frequency of mating how many eggs are worth laying because no more eggs will follow once the male's desire flags.[7]

The eggs are laid as many as three days apart, and each will be incubated for a whole month. During the long weeks of incubation, the female has to get up every twenty minutes and carefully turn the eggs with the underside of her beak; she resumes her place on the clutch facing the opposite direction. When a gentle twittering is heard through the shell, it initiates a "conversation" between the female and the oldest chick after its four-week-long existence in the continuously warmed egg. Two days later, the chick begins the arduous task of sawing holes on the egg world's Tropic of Cancer with its egg tooth; its mother carefully carries away the small fragments of shell in her beak. In spite of her help, it takes at least a whole day for the tiny creature to heave itself out of the shell, the two halves of which are placed next to the nest hollow unless the owl eats them to get needed calcium.[8]

The blind chick barely weighs half an ounce, and its wrinkled pink skin is sparsely covered with a thin, white, downy shirt. If it didn't chirp, its mother could mistake it in the dark for a shrew that happened to be lying around the nest site. The moment the owl gets up to turn the eggs again, the chick's tremulous twittering starts up, clearly a plea for the maternal warmth of the brood patch. Uttering a soft *kht* and then a louder snoring call, it begs for the little mouse nibblies it will take from the mother's beak when she bends down to her chick, chattering gently.

When it's had its fill, the snoring song stops. Before it slides back under the brood patch on its pot belly, its beak is cleaned off, its downy shirt cleaned up. The mother owl keeps the nest hollow clean and dry. She swallows the chick's feces for the first thirteen days of its life; after that, the chick works its way backward out of the hollow and squirts its excrement out with a wiggle of its rear end.

A crack opens up in the slit between its eyelids, ending its eight-day blindness. A few days later the large eyes of this tiny owlet can already peer into the twilight world around it. This is when its second, whitish, downy coat begins to sprout, and in sixteen days it will enclose the chick in a thick fleece from top to bottom.

One chick after another hatches in a rhythm synchronized with the owl's egg-laying. For an average flock of chicks Mr. Owl has to slay dozens of mice every night. The female uses just the guts and tender muscles for the nibblies she passes on. She is careful to distribute food equitably, to clean the nestlings' downy coats or the fleecy suits of the larger ones, to remove feces from the small ones, to keep them warm while simultaneously incubating, turning the remaining eggs, and keeping the hollow clean. The larger nestlings meanwhile cannot fit under the brood patch anymore, so they cuddle up to their mother's plumage on the outside. In the meantime, the female still finds time amid her flock of young to copulate with her assiduous hunter after he has delivered any kind of prey whatever.

At three weeks, the older nestlings are able to swallow a mouse whole, skin and all, and to keep their little siblings warm; the female can fly off now and then to go hunting. While she's away, the bigger nestlings take the chirping little ones into their midst, forming a pyramid of fleece. They have been lying on their stomach or squatting on the calluses of their heels for eighteen days. Now they can stand up, having already taken their first steps. But why just walk when you can hop and flap around? At three weeks they can even stand up and preen their feathers without falling over, and they like to ruffle the edge of their extremely busy mother's facial disk with their beaks. Now their own disks begin to sprout, and their primaries, too. Until their facial disk is fully formed, their long, pointed, still uncovered bird faces will poke out of their downy fleece.

When either parent comes to the nest bearing prey, the big and little chicks begin a snoring contest, but when the adults are away, there is only one young bird that takes care of the beg-and-snore song. The mouse will not be given to the most agitated, pushiest chick, nor to the loudest snorer; parents seem to know somehow which of their children is the hungriest.[9] At one month, the half-grown owlets imitate a chattering sound while eating—the same sound the male emits when presenting his mouse and the female when feeding the chicks their nibblies. Their *gigigigigigi* sounds so promising that it lures the little ones over. If the chattering chick has had enough, it allows the prey to be snatched away. If there is a good supply of food, the big chicks gradually take over some of the parental care and feed their siblings. If there's a dearth of mice, the littlest ones get short shrift, wither away, and are ultimately eaten by their siblings.

In years when the mouse population experiences a massive increase, the female can only hunt for a short time because the male is yet again put to the test by his once-more dominant spouse before a new clutch can be laid. The contentious courtship preceding an overlapping brood will last almost as long as that of the engagement period. The male must conquer the stronger owl once more by performing his dance and presenting his love gifts at a new nest site. To top it off, he must take care of the first young birds all by himself—they were fledged at two months but are still not independent a month later. If this turns out to be too much for him, it can happen that the female looks around for another father for her second brood. But if there are mice galore, many a male can even keep pleasing two owls in two nests at the same time, whereas in years of

Pierre Belon, *Barn Owl*, 1555

famine it may happen that two males will provide for one female because one male cannot come up with enough sustenance all by himself.

In former times, a tremendous explosion of field mice every three or four years used to transform fields and meadows into a Barn Owl Land of Cockaigne. Modern agricultural practices have put a stop to that. Since the 1960s, nobody has needed any advice about how to prevent cyclical plagues of mice. Constant fertilizing makes meadow grasses grow too densely, rendering them uninhabitable for mice. Stubble is immediately ploughed under after the harvest; deep furrows in winter destroy whatever rodent nests are left; there are no longer any embankments, fencerows, or hedges in consolidated farmlands to provide winter shelter. Pesticides have a delayed effect until they eventually poison rats and moles, which become toxic prey for owls and raptors.

In times of field-mouse cycles, there was enough food for all mouse eaters, be they ermine or domestic cat, hawk or owl. When mice are scarce, wild animal populations decline, but domestic cats, fed as they are by humans, will increase in number as always. Shrews are welcomed by owls when fat prey is in short supply, but cats don't relish them. They will kill them and leave them, hunting just for the sake of hunting, but field mice and moles are delicacies for them, too. In this way, they pose another threat to the Barn Owl, which can hardly find enough food anyway to lay large clutches to compensate for a tough, deadly winter.[10]

The modern farmer no longer nails a Barn Owl to the barn door to ward off evil spirits. You might still come across a grandmother here and there who will tell her grandchildren that once upon a time "church owls" would sip oil from church lamps. Nobody nowadays thinks that "snoring owls" are in league with dark powers, and even the emblematic wise old owl is disappearing. But the knowledge that these birds are very beneficial in mouse control seems to have declined along with superstition. You can tell by the church towers in every village. No dove droppings shall befoul any newly restored church. And so screens of wire mesh are put in front of the belfries' acoustic shutters, and "church owls" are deprived of breeding sites safe from martens, sites that used to be called "owl heaven."

Owls were simply able to ignore clanging bells, which meant nothing in their world; but if they do not notice fast-approaching cars as they hunt along the shoulder of a road, it will cost countless male owls their life. Left behind most of the time are a "wailing owl" and her brood. If she cannot attract a new male with her shrieking and test him and love him as her dancing mouse provider, then her offspring are threatened with starvation.

A vent in the gable end of a barn left there specifically for Barn Owls used to be called an "owl levy" (*Eulengebühr*) back when people still knew how indebted they were to these birds.

CHAPTER 9

The Intrepid Hawk Owl

THE FEARLESSNESS the (Northern) Hawk Owl shows toward its foremost foe comes from the isolation of the northern taiga's swamps and conifers, where it rarely sees a human. *Surnia ulula* inhabits the boreal zone in the Old and New Worlds, from Scandinavia to Siberia, from the Tien Shan Mountains to Kamchatka and Sakhalin, from Labrador through British Columbia to Alaska.[1] This vole predator breeds and hunts in open woodland of dark spruce and Arolla pine or in sparkling larch (or tamarack) and birch bordering peat bogs and clearcuts. As an owl highly susceptible to wanderlust, it moves south from Scandinavia when the rodent population crashes in winter; some years it will turn up in Poland, Denmark, and Schleswig-Holstein. An owl banded in Sweden flew more than eleven hundred miles all the way to Perm in the Urals; others have been vagrants to England.

A small pair of owl eyes with sulfur-yellow irises shines from out the oatmeal-colored face framed by a feathered cowl with a brownish-black border. The blurred horizontal band above and across its barred, silky vest is like a fake collar over its breast. The blotches and bands on wing and tail look like sunny postmarks on shaded bark. The Hawk Owl's coloration and marking on its smooth, glossy feathers show it has

Anita Albus, *Hawk Owl (Surnia ulula)*,
in Birch Forest, oil on copper, n.d.

Anita Albus, *Dorsal View of Hawk Owl Skin*,
watercolor and body color on paper, n.d.

adapted to its favorite tree, the birch. Its polelike pose—all owl species use it for camouflage—on a broken birch trunk renders it invisible in broad daylight. Its convex facial disk is cracked open just a bit; its feathers lie flat from furred toes to dotted brow; and there are two rounded corners on the side of its head, like ears. When crouched on a branch in a relaxed posture, it covers its thickly feathered claws beneath a puffed-up silken vest.

The Hawk Owl can fool our eye superbly even without mimicking a birch tree. It flies like a harrier, calls like a kestrel, and hunts the way a shrike does. The slim, long-tailed silhouette, the pointed wings, the rather stiff feathering, and the relatively small head do not fit the typical owl pattern but suggest a raptor, which gives the Hawk Owl its second name, "owl-falcon."

The serrations on the border of the primaries—the outer flight feathers on the wings—create the soundless flight of a nocturnal owl, but the feather edging in the diurnal Hawk Owl is not elaborated in the same way. *Surnia* can afford to make a faint rustling sound because it hunts by day and at dusk and rests during the white nights of the northern summer. It looks for rodents while lurking on a scrawny tree and drops down on them in an almost vertical dive. Sometimes it skims low over the ground, hovering on beating wings as it scouts for a vole; or it darts boldly after a fleeing bird, following its spiraling escape in a twisting flight, then grasping the bird in its claws and locking onto it, somersaulting into the snow, falling all over itself until it adopts the typical owl mantling position by spreading its wings over its prey, then shaking and biting it, then killing and plucking it before finally devouring it.[2]

If a person happens to show up in its forest solitude, *Surnia* doesn't seem to be disturbed. Brehm, an old admirer of the Hawk Owl, says that it "calmly stares at everything, its gaze thereby appearing rather sheepish but sly; but it does not cross its mind to look its most dangerous adversary straight in the eye."[3] It is not even bothered if you toss something at it; it swivels its head 180 degrees and regards the passing objects with astonishment. But any two-legged invader who tries to peer into its nest or approach one of its nestlings perched on a branch will be plucked at like a Red Grouse that the owl feeds on in winter. If wounded and cornered, it puts its back to the wall and fights for dear life.

Surnia's exceptional, ocarina-like ululation is softer and less eerie than a nocturnal owl's call. The territorial song of the male "owl-falcon" in spring begins with a softly tripping trill, rises to a rolling trill, and swells into a long, floating vibrato: *hu hu hu ee-ee-ee-ee-ee* . . . [4] The tail feathers wag in the same rhythm. The richly modulated song of the unpaired female owl sounds somewhat brighter and is more wavering. The male Hawk Owl has to tame the reluctant female before the wedding, as do all owls. Feathers fly until the female chosen as that season's mate is ready. If his intended's fury still hasn't subsided after frantic chases, fierce billing, and wild claw-to-claw fights, she vents her fury by battering and squeezing some dead prey, or else she digs her claws into a branch and flaps her wings. When the male, after repeatedly visiting a nesting site in a tree hollow, has finally persuaded her by chirps, moans, and groans and has won her heart with melodious trills and lavish gifts of mice, they sing a trilling duet together on a branch, *con fuoco*, followed with avian swiftness by copulation. Feather ruffling ensues once more.

Wolfgang Scherzinger observed Hawk Owls copulating in his compound only once or twice an evening. Nonetheless, almost all the five or six eggs in each clutch were fertilized, having been laid two to three days apart. The male Hawk Owl provides for his mate; she is continuously on the nest and will beg for food with tail cocked, body aquiver, head gyrating, and a call like a nestling's. Her monosyllabic *chyeet* or disyllabic *gyeeee-ih* sometimes sounds hoarse and drawn out, at other times like a whistled squeak interspersed with trills and whinnies.[5]

If an interloper attempts to climb up to the nest site, he must reckon with an enraged Hawk Owl dive-bombing him with a piercing staccato alarm call, *rikrikrik* or *zikzikzik*, ripping his hat off and tearing out clumps

Hawk Owl.

John James Aububon, *Hawk Owls*, pl. CCCLXXVIII, 1827–38

John Gerrard Keulemans, *Pair of Hawk Owls*, 1897–1905

of hair. If the intruder still hasn't given up trying to inspect the nest hollow, the incubating owl will keep sitting on her eggs, stock-still and silent, when the trespasser's wounded head appears in the entrance hole.

The wee chicks seem uncommonly fragile when, after four weeks, they hatch a day and a half apart from their spindle-shaped, rather dull eggs. They weigh one- to two-thirds of an ounce. Their first downy coat is thin, white, and plush, providing sparse cover for their sensitive pink skin. These blind, delicate creatures lie flat on their stomach for the first three days. Their mother arranges them in a circle around the remaining eggs and shelters them tirelessly. With a hollow, stuttering *yuyu-byu-byu* she serves them up meaty morsels from the voles the male has brought to the nest-hole.[6] The little ones can sit up after three days and trip around on their heels a little. After they can stand, at around eight to ten days, and learn to dance around on their toes, barblike quills pop out of their second, darker down feathers. With their eternal *pchee-yet* or *chee-epp* they beg with a monosyllabic *pfiekh* or disyllabic *pfee-yee* for the vole shish kebab they've been craving. They cannot gobble down undissected voles until they are two weeks old and their eyes are open. When the young leave the nest at three and a half weeks, they still have their fluffy, bright, gray-and-brown second down feathers and are not yet fully fledged. They climb adroitly around the branches, pulling themselves up tree trunks by hooking their beaks into the bark; they flutter and hop when they move around on the ground. As soon as the nestlings are fledged, at thirty or forty days, they beg their parents for prey by swiveling their heads and quivering while giving a squeaky, hoarse crescendo call to signal their hunger. When they have reached three months, their spotted, streaked, and barred feathering is barely distinguishable from their parents'. They are as lively and boisterous as the adults, and their jerky, lightning-quick movements make them look edgy, even frantic, just like the old folks.

A bath in water or snow delights *Surnia ulula* as much as collecting the sun's rays on its outspread feathers. Every Hawk Owl, every

Christian Buhle, *Hawk Owl*, 1835

"owl-falcon," is a loner and hunts by itself. As vast as the taiga is, the bird's habitat has shrunk, and the species has declined appreciably since the nineteenth century. Until now, the Hawk Owl has not been red listed. It still flies, largely unobserved, in the long twilight time of the northern hemisphere, now gliding, now diving, now beating its wings as it hovers over peat bog and clear-cut.[7]

CHAPTER 10

The Wise Kingfisher

───────

THE GREEK legend that gave *Alcedo* its name is as stirring as the bird's beauty. The gods took pity on the faithful Alcyone, the daughter of Aeolus, the god of the winds, and transformed her and Ceyx, her deceased husband, into kingfishers. In Ovid's retelling:

> The Gods their Shapes to Winter-Birds translate,
> But both obnoxious to their former Fate.
> Their conjugal Affection still is ty'd,
> And still the mournful Race is multiply'd:
> They bill, they tread; Alcyonè compress'd[1]
> Sev'n Days sits brooding on her floating Nest:
> A wintry Queen: Her Sire at length is kind,
> Calms ev'ry Storm, and hushes ev'ry Wind;
> Prepares his Empire for his Daughter's Ease,
> And for his hatching Nephews smooths the Seas.[2]

There is a legend that the Kingfisher's nest floats on the sea, a fable harking back to Aristotle. We read in Book IX of his *History of Animals* that the nests are pale red and shaped like a long-necked gourd larger in size than the largest sponge; that they have a roof, with extensive solid

Johann Theodor Susemihl, *The Common Kingfisher*
(*Alcedo ispida*), gouache, ca. 1800; reproduced with the kind
permission of the Hessisches Landesmuseum, Darmstadt

and hollow parts unable to be cut even by the sharpest knife; but that they can shatter like sea-foam (i.e., meerschaum, a leather coral) when crushed with one's hands. Its narrow mouth, serving as an entranceway, is located in such a way that even in rough seas water cannot get in. And, Aristotle concludes, there is a question as to the material *Alcyon* constructs its nest with: it is thought to be largely made from bones of the needlefish (*Belone belone*), for these birds live by eating fish.[3]

Pliny writes in much the same vein:

It is a very great chaunce to see one of these Halcyones, and never are they seene but about the setting of the starre Virgiliae, [i.e., the brood hen] or else neere mid-summer or mid-winter: for otherwhiles they will flie about a ship, but soone are they gone againe and hidden. They lay and sit about mid-winter when daies be shortest: and the time whiles they are broodie, is called the Halcyon daies: for during that season, the sea is calme and navigable, especially in the coast of Sicilie. In other ports also the sea is not so boisterous, but more quiet than at other times: but surely the Sicilian sea is very gentle, both in the streights and also in the open Ocean. Now about seven daies before mid-winter, that is to say, in the beginning of December, they build; and within as many after, they have hatched. Their nests are wonderously made, in fashion of a round bal: the mouth or entrie thereof standeth somewhat out, and is very narrow, much like unto great spunges. A man cannot cut and pierce their nest, with sword or hatchet; but break they wil with some strong knocke, like as the drie fome of the sea...[4]

In Pliny's time, no one had ever seen a Kingfisher on the nest. But since this diving bird's nest was nowhere to be found among reed beds or riparian bushes, then where might it be if not on the water? A bird with such sparkling, green-blue, and rusty red plumage, a bird that can calm a December sea, merits the most wondrous nest of all. Polyps broken off from a pale red species of leather coral would float on the sea and so were thought to be Kingfisher nests. *Alcyonium* was the name Linnaeus chose for the genus of these eight-armed corals in remembrance of the legend of the Kingfisher's nest.

To Plutarch, this construct, woven together by a bird's bill, seemed disproportionate to its meerschaum-like fragility. As the Kingfisher's greatest admirer, he wrote in his *De sollertia animalium* that he considered its nest nearly indestructible; that it resembled a fisherman's bow-net and was so finely and tightly woven that it could hardly be broken with stone or iron; that no other creature, nor even a drop of water, could find its way inside:

> [The Kingfisher is] the wisest of sea creatures, the most beloved of the gods! For what nightingales are to be compared with the halcyon for its love of sweet sound, or what swallows for its love of offspring, or what doves for its love of its mate, or what bees for its skill in construction? . . .
>
> When the halcyon lays her eggs, about the time of the winter solstice, the god brings the whole sea to rest, without a wave, without a swell. And this the reason why there is no other creature that men love more. Thanks to her they sail the sea without a fear in the dead of winter for seven days and seven nights. For the moment, journey by sea is safer for them than by land. If it is proper to speak briefly of her several virtues, she is so devoted to her mate that she keeps him company, not for a single season, but throughout the year. Yet it is not through wantonness that she admits him to her company, for she never consorts at all with any other male; it is through friendship and affection, as with any lawful wife. When by reason of old age the male becomes too weak and sluggish to keep up with her, she takes the burden on herself, carries him and feeds him, never forsaking, never abandoning

Pierre Belon, *Kingfisher (Halcyonium)*, 1555

him; but mounting him on her own shoulders, she conveys him everywhere she goes and looks after him, abiding with him until the end.[5]

All the tragic elements of Alcyone and Ceyx's tale are preserved and changed dialectically in the Kingfisher's legendary life. Another Alcyone, now a star in the Pleiades, had gone with her sisters up into the heavens, but here on earth there was new life needing shelter in that very element which had swallowed up Ceyx in its fury. His storm-tossed, shipwrecked vessel finds its counterpart in the boatlike Kingfisher's nest, as immune to destruction as was Ceyx and Alcyone's love. Her name comes from *alky* + *on* and means "possessing the gift of warding off evil." So if you had a Kingfisher's skin on your person, this was supposed to shield you from Zeus's lightning bolts. And those leather corals named after the Kingfisher were also ground up for medicines and cosmetics.

Alciquium (ἀλκυών or *alkuōn*) is what the Greeks called the *Halcyonium*, as Pierre Belon tells us in the chapter on the Kingfisher in his *L'Histoire de la nature des oyseaux*, which has a precise description of its true nest. Belon saw nesting holes on the banks of two rivers in Macedonia; these brëeding burrows were "four to five feet deep" and led to a hollowed-out chamber. Delicate fish bones and scales there—correctly deduced as being disgorged pellets—formed a kind of nest. It likewise struck Belon as plausible that the bird would breed in winter. When northern rivers iced up, the birds would presumably fly off to mild Mediterranean climes and apparently reproduce for a second time that same year, during the "halcyon days."[6]

Even Conrad Gesner recognized that the "Eyßvogel" was a hole nester, and Agricola counted it among the "creatures from below ground."[7] Their century, the sixteenth, witnessed the rise of the art of the emblem, which kept classical myths alive. *Miratur natura silens* ("Here nature admires in silence") is an inscription on one of countless kingfisher emblems. Far from the prying and prodding of science, the wise little bird nests on a waveless sea. If a solitary rock provides a platform for its nest, this can do its confidence no harm:

> The noble bird, the Alcyon
> is on a rock, its home /
> in th' sea / its nestes throne /
> unbedousèd by the waves.
> And should it breed / and young it feed:
> Wind / winter / must be silent.
> God doth it save:
> and ev'ry wave
> will ne'er more be defiant.[8]

In the moral world of emblems, almost every animal can be interpreted *in bono* or *in malo*, depending on the context in which it appears. But not so for the Kingfisher. Whether it symbolizes marital fidelity,

Kingfisher Emblem, in Johann Ulrich Krauß,
Les Tapisseries du Roy, 1690

Anita Albus, *Pair of Kingfishers (Alcedo ispida)*
in a Landscape, watercolor and body color on vellum, n.d.

the productivity of peace, farsighted prudence, a sense of *kairos*, leisure time for art or calming the mind, unshakable trust in God, constancy, or justice—the bird with the most resplendently colored feathers in our hemisphere always sets a fine example. Such a magnificent, glamorous creature must have seemed the incarnation of *sophrosyne* to an age for which the Beautiful was unquestionably the Good. A Kingfisher is a Kingfisher—not the same creature as a person called a "kingfisher." That person was considered sly, shrewd, crafty, and scheming.[9]

Gone are the days when moralizing would transfigure fine-looking birds. The Kingfisher is stereotyped today as a "flying jewel." That is why the unsociable "king's fisher" enjoys great popularity as a photographic subject. But no photographer can capture the dynamic beauty of the structural color of its plumage:[10]

> The green and blue colors sparkle and shimmer most splendidly, one quickly changing into another when the angle of the light varies as the bird turns, or when it is seen from a different position and in a different light. In completely bright light, for example, those gorgeous colors blend into a unique, prismatically magnificent blue green, seemingly poured over the entire bird from above. But if the bird is seen in semidarkness, its colors all revert either to a lovely ultramarine or to a marvelous, somewhat darker azure blue, depending on how the light strikes it, from which side, which angle, and so forth. The countless gradations shade from the deepest blue over to the brightest green so that it should come as no surprise if someone calls a color green that someone else takes to be blue at the same time. Hence the diverse designations for the main color in descriptions of the bird.[11]

The little bird is infrequently seen in the wild. The azure-blue and malachite-green plumage of an immobile, hunting Kingfisher is invisible amid the dappled green of a bank mirrored in the glittering water and combined with the blue of the sky. Unless you hide in the

Conrad Gesner, *Kingfisher*, 1617

bushes and lie in wait for a long time, you will only spot it if it shoots, arrowlike, through the air on whirring wings. A turquoise flash, a sharp, clear whistle, and it's gone. The fleeing bird's penetrating, two-syllable *chi-eeeee* can indeed sound like *keyyyx* to a partial listener. On the other hand, the glorification of the Kingfisher's song in antiquity can only be interpreted as a confusion with the beautiful song of the Sedge Warbler (*Acrocephalus schoenobaenus*).

Opinion is divided on the bird's *sophrosyne*. Brehm says: "It is a quick, wild, quarrelsome, shy bird that always lives in solitude and refuses to socialize, particularly with its own kind. One usually observes it in the shade of a bush or some reeds, sitting calmly on a stone, branch, post, or similar location overlooking water as it lies in ambush for fish; yet it never forgets its safety while thus employed. From time to time it cautiously raises its head, peers all around, and flies off as soon as its suspicions are aroused."[12]

Naumann describes the European Kingfisher in much the same way:

Compared with others of its ilk it is such a wrangler that two of them can never bring themselves to live in the same neighborhood, the only exception being during the breeding season if two of them mate and there is plenty of food in a particular place. Otherwise they will chase and nip at each other until one of them gives way and seeks out another territory... One will go after the other like an arrow, skimming the water, cornering around a bank in a very quick turn while constantly calling so that in their blind rage they often do not see someone standing on the bank until they are close by, but then they get altogether so frightened that they forget their quarrel for the moment and fly off in different directions; for they are very shy birds, which in moments of excitement will flee an approaching person when he is still far off, and which therefore dwell in quiet, secluded retreats on a body of water, putting their fierce nature aside a little only when food is scarce or winter days are cold.[13]

At the beginning of the breeding season rival male Kingfishers will engage in long duels of dominant display. The adversaries attempt to intimidate each other by sitting upright, flattening their body feathers, hanging their wings forward, and opening their bills wide. Slow-motion bows interspersed with chases intensify their threats, culminating with outstretched wings held high. Should neither give way after several hours, it may come occasionally to a battle of bills; extreme cases might even end fatally.

The Kingfisher's "mostly" monogamous seasonal marriage for breeding purposes begins in late winter and lasts until September. Pairing commences with much calling and many dashing aerial moves. Sometimes the birds chase each other along a river, sweeping low over the water, sometimes they fly high into the treetops, where they are seen only in the mating season. If the male has found a vertical river bank and

Anita Albus, *Kingfisher (Alcedo ispida) with Trout*, still life
(detail), watercolor and body color on vellum, n.d.

stays there hovering to check it out with his bill for its suitability as a
burrow, the expectant female demonstrates her interest with agitated
calls. Before achieving copulation, *Alcedo* must show his mate his fishing
expertise; with a deferential bow he passes her, from bill to bill, young
brown trout, minnows, bullheads, roaches, bleak, or sticklebacks. Without returning his bow she accepts his wedding gift with wings aflutter.
A Kingfisher will consume around three-quarters of an ounce of fish a
day, about one-half of its body weight. Diving for prey takes but seconds.
The "king's fisher" speeds like an arrow from its perch into the water. As
it plunges in, wings held tightly against its body, it draws its nictitating
membrane over its eyes. Before seizing its prey it slows its impact by putting its legs forward and flapping with deep wing beats. With the fish
in its bill pressed to its breast, it shovels its way back to the surface with
outspread wings. Once above water, it jerks its bill and its prey upward,

showering droplets all around before darting back to its perch. It kills the fish by banging it repeatedly against a stone, branch, or pillar, unless it's small enough to squash with its bill and achieve the same end.

The male begins building the nesting tunnel by hovering and pecking earth from the bank with his daggerlike bill. Upon returning to his mate nearby, he shakes the dirt from his bill and immediately resumes his hovering and picking away. The female soon takes over, the bridegroom standing on guard as she digs. Not until a landing area is created that lets the bird stand while working does the work ease up by half. It busily kicks the dug-out dirt behind it with its hardworking little feet, clearing out "construction waste" by flicking it backwards with its short, round, spade-shaped tail. It takes ten to fourteen days to create a gently upward-sloping tunnel, just over three inches high and fifteen to thirty deep, leading to a five-inch high chamber almost seven inches across.

Bathing and courtship feeding will interrupt the birds' construction project. The only way to get out is to walk backwards because there won't be any room to turn around until the chamber is hollowed out. Either bird will appear tail-first in the entrance hole, take off backwards, make an immediate about-face, and plummet into the water bill first to take a cleansing bath. After a number of dives it lands back on its perch and thoroughly shakes the water off its feathers. Then it meticulously grooms itself with its bill, feather by feather, and rubs its head on its back, breast, and the bend of its wings, gnawing its toes, and oiling all its feathers with a secretion from its preening glands. A wing stretch and a yawn conclude its toilette.

When the burrow is not quite finished, the Kingfishers' wedding takes place. The groom, after presenting his gift of a fish, assumes his dominant display, which resembles his threatening posture, and points his bill to the sky. The bride snuggles down on a branch, horizontally, her tail pointed up to the side to invite the bridegroom to copulate as she emits a few calls. He immediately flies at her from behind, drops onto her back, hovers, and holds her neck feathers with his bill during the brisk copulation. Her spouse subsequently favors a bath, whereas she is typically content with shaking her ruffled feathers smooth.

Copulation, an event repeated several times a day, is not always preceded by courtship feeding. Whether Mrs. Kingfisher is fed or dives for fish herself, she will disgorge oval-shaped pellets in the burrow's chamber. If a tunnel is used several times, a thick layer of white, silver-sparkling fish bones and scales will cover the chamber floor right up to the walls. If you find a pellet of bone or scales beneath a Kingfisher's lofty perch, it will easily crumble under your touch. The female almost invariably lays one egg a day for seven days. The eggs are approximately seven-eighths of an inch by seven-tenths and weigh about one-seventh of an ounce; they are comparatively large for her petite body, which is not much bigger than a House Sparrow's (*Passer domesticus*). In the earliest stage the reddish yolk can be seen shimmering behind the finely pored shell. Perfectly smooth, glossy, and as white as precious porcelain, the not quite round eggs are as beautiful as the bird they emerged from.

Incubation takes nineteen to twenty-one days and begins when the sixth or seventh egg is laid in the thick layer of fish bones if the nest is old, or in the skimpily padded hollow of a new one. *Alcyon* and *Alcyone* alternate at this time even though the female is more often called upon and usually takes over the nighttime brooding herself. This is when the birds must keep as quiet as possible; they otherwise call quite frequently. An incubating bird faces the light at the end of the passage and will exit the hole after hearing a short *chee* from the bird flying up to the steep bank to relieve it.

A three-week incubation period—and then, every few hours, one Kingfisher chick after another hatches out of its glossy globe. The pink nestlings, naked and blind, lie like a heap of worms beneath the feathery hood of whichever adult happens to be on the nest at the time. The adults feed them in rotating flights. The bringer of a little fish announces

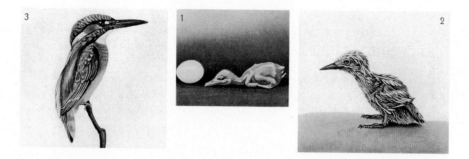

Oskar Heinroth, *Kingfisher (Alceda ispida): 1. naked nestling with egg;
2. 9 days before fledging: 3. adult;* vol. I, 1926–33, colored photographs

itself to the next generation with a hoarse *kraay,* and is answered by a
whispered *vree,* which soon intensifies into a penetrating, persistent
rreeuhreeuhreeuh. It's usually just a single nestling, the one who's next in
line for food, who will be singing for its supper. The Kingfisher was a
good choice for a symbol of justice. The chicks form a rosette in the darkness of the chamber; each places its breast on the back of the bird in front
and turns its heavy head sideways to face the center, and so a circle is
formed. The nestling lying nearest the tunnel is the only one with its head
turned to the light. The moment the entrance hole is darkened, it opens
its beak, takes the fish tendered lengthwise and head first by the waddling
adult, and swallows it down; the tail of an occasional larger specimen will
stick out for a while longer. Having had its fill, the chick turns around
and squirts some thin, watery excrement toward the light at the tunnel
entrance. Whereupon the Kingfisher carousel shifts one notch ahead,
and the next bird in line occupies the coveted feeding station.

Oskar Heinroth once adopted six Kingfisher nestlings. His foster
children lived in a dark flower pot with a shard broken out to make a
hole for a paper tube through which the young could be fed small fish
with a pair of tongs. When the baby birds were first brought to Heinroth, their eyes were "already fairly well open, but their eyesight was poor,

and they did blink a lot."[14] At that age, about twelve days, they are still wearing a colorful, prickly coat. Every feather of their future stunning plumage—blue, green, rusty red, or white—is still completely encased in a translucent covering that falls off at three weeks. Seen through this dull horn casing, the feathers, which will eventually unfold in iridescent blue and green hues, appear as a uniform light blue, the rusty red feathers isabelline in color.

Heinroth was the first to observe and describe the Kingfisher carousel:

As soon as we had placed the hedgehog-birds into their artificial nest, they immediately organized themselves in the position shown in the black-and-white photographs. Most of the time one of them will lie with its beak in the opening, waiting to have its parents or tongs deliver its fish. Then it moves clockwise, to the left, and the next one replaces it at the opening: that is how each receives its due ... This answers the question in a single stroke: how is it that not one of the six still blind young is forgotten in that dark, underground den, where the adults themselves cannot see properly. The same procedure is followed when the chicks void ... We discovered that it was always the young bird sitting at the opening that would turn and squirt its stream of excrement along the passageway, which would soak into the clay bank outside the chamber, between the entrance hole and the nest, seeping at least partly into the permeable soil.

The feces would squirt far beyond the short paper tube and onto a blotter. "You would never on your life look into that hole unless you wanted to risk getting a free gift in the face."[15]

Kingfishers slope the tunnel from the entrance hole to the brooding cavity at a fifteen- to thirty-degree angle upward, which lets the feces flow downhill. Science is baffled as to how the nestlings are able to stand the evaporation in their cesspool without harm. The layers of feces and the rotting fish leftovers give off high concentrations of carbon dioxide

John Gerrard Keulemans, *Kingfishers (Alcedo ispida)*,
adult male and two juveniles, 1897–1905

and ammonium. The stink presumably keeps mice and the Least Weasel (*Mustela nivalis*) from going up the slippery slope to get at those delicious chicks. Adult Kingfishers dive into the water whenever they leave the tunnel. Two weeks after the offspring have hatched, the parents only enter the burrow to feed them. The potbellied hedgehog-chicks (underground chicks) crowd together for warmth. They now overlap like roof tiles, beaks toward the chamber entrance, and whichever wants the next fish will run toward the fish bearer, trilling loudly. As nesting time nears its end, a pushy chick will sometimes jump the line, and then the whole gang gets some fighting practice.

When their dapper feathers burst out of their "prickles" after the third week, the nestlings give them a thorough cleaning, using their beaks to remove any horn casings left hanging. Little Kingfishers, looking just like the grown-ups to a feather, now fill the entire brooding chamber. A *chick* or *tsipp*—Heinroth says it sounds like a bird sneezing—heralds their imminent exit from the burrow a few days later. Early one morning, one Kingfisher after another whirs out of the dark into the light, either under their own steam or egged on by the adults.

Zipping straight ahead goes well, though the chicks can't yet hold their stumpy-tailed stern horizontal. It's not until their wing and tail feathers are fully grown that the young will learn how to maneuver and land steadily on their little feet like an adult bird. For the time being they don't often hit the branch they're aiming for, so they flutter and flounder through the bushes. They will still be fed for two or three days. To their ears, *teet* means "Here comes a fish!" and their answer, *chick*, means, "I'm over here!"—the greenery makes them hard to spot, even for an adult.

When the time for parenting has passed, the order is "Go fish for yourself." It's the diving that is especially difficult for such a small, lightweight bird. Unless it plunges down with all its might, it will float like a shuttlecock. Once it gets the hang of submerging, it has to learn to tell a fish from a drifting leaf. If it has grabbed a leaf it will beat it against a branch, but try as it may, that leaf will never turn into a fish . . .[16]

It's bad enough that the parents don't bother themselves with the new generation anymore; adding insult to injury, they also drive them out of the nesting territory. Every young "king's fisher," male and female, must head out to conquer its own fisher kingdom. The siblings disperse in all directions in their hunt for a territory. Many fledglings will wander just a few miles away; others will put more than a thousand miles behind them. Females will usually travel farthest, whereas males tend to find their fishing grounds and future breeding territories closer to their home neighborhood.

The parents keep themselves busy in the meantime with a second brood. A brief courtship has preceded the second nest of eggs, now laid in a different breeding burrow. If two broods overlap, the female will incubate by herself while the male keeps feeding the first brood and providing for his mate on the nest. He has to supply around fifty-six finger-length fish daily for seven three-week-old nestlings. Monogamy would seem advisable. Any Kingfisher having two mates simultaneously must carry out double the feeding and flying between territories.

The parents can recommence their rotating incubation as soon as their first young have been expelled from the breeding territory. The second brood moves out of the burrow in mid-July or the beginning of August, depending on when incubation began. Very few Kingfishers make it to a third brood, which would fledge sometime between the end of August and the end of September; an extremely rare fourth brood would not be out of the nest until October. Ornithologists have observed that if there are two or three broods, the young of that same year will occasionally enter burrows with begging nestlings to assist with the feeding unless the adults shoo them off. Studies have shown that these helpers, against all expectation, were not related to the birds being fed.[17]

Not every incubation ends happily. A new generation will thrive if the weather cooperates. Fishing is bad in murky waters. Prey is invisible after a cloudburst until the disturbed clay settles. Even a drizzle will muddy the water surface and make hunting difficult. The smaller the nestlings, the more quickly they will starve to death. High water may flood out nesting burrows. Foxes, badgers, raccoons, wild boar, martens, etc. might dig out a burrow and down the eggs and chicks. Photographing underground hedgehog idylls costs many broods their lives because photographers dig up the nesting chamber.[18] Other nestlings perish when hordes of surfing, swimming, barbequing, or rampaging bipeds prevent fishing and feeding for long hours on a sunny weekend.

Alcedo's life is short: "About 80 percent of Kingfisher fledglings will die between the time they leave their burrow and the next breeding season, and around 70 percent of adults will die in the course of a year. All but a few will live no more than three years, and it is exceptional for a bird to survive its fifth year."[19] Long, icy winters spell death for the Kingfisher. Its fat reserves, stored up from November on, are barely sufficient for it to carry on after two days without food. Since Western Europe sees a hard winter only every eight to ten years, the Kingfisher remains loyal to us during the cold season. It's only in Eastern Europe that they are forced to go south to avoid long-lasting frosts.

Because the bird is protected, Kingfisher feathers no longer adorn ladies' hats, and anglers have to make do with duck, partridge, Golden Pheasant, or Marabou Stork feathers for tying flies. *Rare* Kingfisher species can now live in peace even with fish farmers, their enemies in bygone days. Persecution is no longer a direct threat but ruthless destruction of the bird's breeding areas is. *Alcedo*'s world includes unchannelized rivers meandering through wild meadowland; streams bordered by willows weaving through secluded valleys; still waters in woodlands free of jogging trails; clean fishing grounds rich in different species; and lakes virtually toxin free, their banks inaccessible to recreation fanatics. If they cannot find a steep bank, some Kingfishers will search for a breeding station far from water, deep in the woods, and burrow into the root balls of huge fallen trees.[20]

The Kingfisher family is divided into two sub-families comprising fourteen genera and twenty-three species. *Alcedo atthis*, the Common,

Anita Albus, *Still Life with White-throated Kingfisher
Skin* (*Halcyon smyrnensis*), gouache on vellum, n.d.

or Eurasian, Kingfisher, is one of twelve species in the genus *Alcedo* and occurs from Eurasia and North Africa to Sakhalin, Japan, Sri Lanka, Sumatra, and the Solomon Islands. The nominate form, *Alcedo atthis atthis*, is found in the Mediterranean area, in the Maghreb, the Middle East up to Northwest India, and Russia as far south as Lake Baikal. *Alcedo atthis bengalensis*, a somewhat smaller form showing a touch more blue, extends from the northern part of the Indian subcontinent and northeast Asia into eastern Siberia and the eastern limit of the Kingfisher's range. *Alcedo atthis taprobana* lives in southern India and Sri Lanka; three island races are indigenous to Indonesia and Melanesia. Europe's Common Kingfisher, *Alcedo atthis ispida*,[21] has an empire reaching from the British Isles in the West to Belarus in the East, and from southern Sweden in the North down to Serbia, Bosnia and Herzegovina, northern Italy, southern France, and northern Spain in the South.[22]

On our smallest continent lives the largest Kingfisher of all—the Kookaburra, Australia's unofficial national bird. Of the four species found there, the Laughing Kookaburra (*Dacelo novaeguineae*) is the best known. Its habitat is open woodlands, where it nests in hollow tree trunks or in tunnels built in large termite mounds. It wouldn't dream of driving its grown-up young out of its territory because it has brought them up so well. The offspring defend the territory together with their parents; they also help with brooding the eggs and feeding their younger siblings. The Kookaburra hunts for large insects, lizards, crabs, and small birds. It loves snakes, too; if it can't kill one by stabbing it with its bill, the bird will take it up in the air and drop it onto the ground until the snake surrenders.

This snake exterminator was a sacred bird for the Australian Aborigines. Its ringing, laughing song, heard most often at sunrise and sunset, was their chronometer. But to the white man's ear, its wild cries of joy sound like the harsh, scornful laugh of a demon filled with schadenfreude, which first earned it the name "Laughing Jackass." In Germany,

John Gerrard Keulemans, *White-throated Kingfishers*
(*Halcyon smyrnensis*), adult male, in foreground, with juvenile, 1897–1905

John Gould, *Laughing Kookaburras*
(*Dacelo novaeguineae*), lithograph, 1840–48

it's called "Laughing Jack" (*Lachender Hans*), "Hunting Kingfisher" (*Jägerliest*), and "Giant Fisherman" (*Riesen- fischer*). The name the Aborigines gave to their sacred bird would eventually gain worldwide acceptance.

The Laughing Kookaburra's scientific name has wavered between *Dacelo gigas* (or *gigantea*), "Giant Dacelo," and *Dacelo novaeguineae*, "New Guinea Dacelo." The genus name "Dacelo" is an anagram of alcedo. Pieter Boddaert christened it with the species name *gigas* in 1783. That same year, Johann Herrmann named it *novaeguineae* in the erroneous belief that the species was found in New Guinea: Pierre Sonnerat, after a voyage to the Moluccas from 1769 to 1772, brought back a Laughing Kookaburra skin, fraudulently claiming he'd obtained it on mainland New Guinea, when in fact he had never been there. For a long time, the Laughing Kingfisher bore the name *Dacelo gigas*, but since whoever publishes a scientific name *first* has the honor of being the "godfather," the bird is classified today as *Dacelo novaeguineae*. Because Hermann's publication appeared a month before Boddaert's, ornithologists have now declared Hermann's designation the winner. In reality, the Laughing Kookaburra is the only one of the four Australian Kookaburras that does not occur in New Guinea. It must have had a good laugh over all this.

Naumann lists the White-throated Kingfisher, *Halcyon smyrnensis*, among the birds of Central Europe because an ornithologist once sighted it on the Ems River in northern Germany; it is indigenous to the Middle East and Asia and almost twice the size of the Common Kingfisher.[23] The North American Belted Kingfisher, *Megaceryle alcyon*, finds a place as an equally rare accidental among Central European birds in Glutz von Blotzheim's monumental work.[24] Compared with *Alcedo atthis*, the Belted Kingfisher, which ranges to the northern tree line, is a giant. Its wings and tail are twice as long; its weight is three or four times as much as our little "king's fishers." It is distinguished by a slate-blue crest and black quills; the male's white band on the back of his neck contrasts with the gray-blue feathers on his upper side and terminates in a white bib on his throat and chin, and a gray-blue breast band graces his white underside. The adult female has a gray-blue band through the white of her breast as well, but she also has a cinnamon belt that gives the species its common name.

Belted Kingfishers breeding in the cold North will migrate south in winter, as do Common Kingfishers in Europe, whereas the rest of them will either be residents or visitants. The Belted Kingfisher's flight is slower than *Alcedo*'s, and it travels longer distances at high altitudes, alternately flapping and gliding. Audubon describes a Belted Kingfisher that hovered "in the air, like a Sparrow-hawk or Kestril [*sic*], and inspect[ed] the water beneath, to discover whether there may be fishes in it suitable to its taste. Should it find this to be the case, it continues poised for a few seconds, dashes spirally headlong into the water, seizes a fish, and alights on the nearest tree or stump, where it swallows its prey in a moment."[25]

High-speed hunters such as kingfishers, hawks, swifts, hummingbirds, and parrots have two foveae (small depressions in the retina) in each eye, allowing their eyes to focus sharply on their prey and estimate distances better. Swallows and terns are equipped with three foveae in each eye.

The Belted Kingfisher's life is similar to that of the diminutive Common Kingfisher. The prickly nestlings in North America admittedly have a better sense of hygiene than European "hedgehog birds." They peck around on the burrow walls so that their excrement gets covered and absorbed by dirt crumbling off the walls. Nestlings reared by hand and kept in cardboard boxes will bang against a box for as long as it takes to have their befouled litter replaced.

Belted Kingfisher parents will take turns incubating, as they do in Europe, and the female will likewise sit on the eggs at night, when the male is sleeping high up in a tree in the nearby woods, preferably at the top of a conifer. Monogamous mating throughout the season is the rule.

John James Audubon, *Belted Kingfishers*
(Megaceryle alcyon), pl. LXXVII, 1827–38

Anita Albus, *Kingfisher (Alcedo ispida) in a Miniature*,
watercolor and body color on vellum, n.d.

Belted Kingfishers are unsociable before and after courtship and breeding, just like their European cousins. Occasional group hunting by five
or six birds has only been observed during the spring mating season. It
would never occur to a Belted Kingfisher to share his territory with any
of his own kind.

The Belted Kingfisher is even reflected in an instructive Iroquois legend. It takes place in the days when man and beast could converse with
one another and morph into other creatures at will.[26]

KINGFISHER AND HIS NEPHEW

An old man and his nephew were living together in a good home
near the river, where they enjoyed themselves day after day. One
morning the old man said to his nephew, "When you are a man,
remember in hunting never to go west; always go to the east."

The young man reflected and said to himself, "Why should
this be so? My uncle To-bé-se-ne always goes west and brings
home plenty of fish. Why should he tell me not to go? Why does
he never take me with him?"

He made up his mind at last that he would go, ignoring the
advice. So he set off in a roundabout way, and as he passed the
marsh land near the river he saw his uncle. "Ha!" he thought.
"Now I know where he catches his fish." And he watched his uncle
take two sharp sticks from his pocket and put them in his nose,
plunge into deep water, and come up with a nice fish. He watched

him carefully and returned home. The uncle came back with some nice fish but never guessed that his nephew had seen him.

The young man now felt certain that he could fish as well as his uncle. One day, when the old man had gone deer-hunting, the young man thought it a good opportunity to try the new method. He hunted among his uncle's things until he found two sticks, then set off to the same log where he had seen his uncle sitting, which projected above the river. When he saw the fish swimming about, he at once stuck the two sticks into his nose and plunged into the water. The sticks went in deep, making his nose ache dreadfully, and he felt very sick. He hurried home and lay down, thinking he should die of the agony. When his uncle came home and heard him groaning, he said, "What ails you? Are you sick?" "Yes, uncle," replied he, "I think I shall die. My head is sore and pains me." "What have you been about?" asked the uncle severely. "I have been fishing," confessed the young man. "I took your things, and I know I have done wrong." "You have done very wrong," said the uncle; but he took some pincers and drew out the sticks, and the young man promised never again to fish in the west, and soon he got well.

After a while, however, he thought that he would try once more, though he had been forbidden. So he started west. He heard boys laughing, and because he had none to play with, he joined them. They invited him to swim with them and he accepted, and they had a very gay time together. At last they said, "It is time to go home; you go, too." Then he saw that they had wings, and they gave him a pair and said, "There is an island where all is lovely; you have never been up there—over the tall tree, up in the air; come." So they started up in the air, far away above the trees till they could see both sides of the river; and the young man felt very happy. "Now," said they, "you can see the island"; and he looked down and saw the print of their tracks on the island; so he knew they had been there. "Let us go in swimming again," they said. And they went into the water. "Let us see which can go down and come up the farthest." And they tried one at a time, and the young man was last, so he had to go the farthest. While he was in the water the rest put on their wings, so he put his on too and flew up into the air. He pleaded in vain for them to wait; but they called, as though speaking to some one else, "Uncle, here is game for you tonight." They flew away in spite of his entreaties, and he thought to himself, "I shall surely be destroyed, perhaps by some animal."

As he looked around he saw claw marks of dogs scratched on trees and concluded that perhaps they would tear him to pieces. To confuse them in tracking his scent, he climbed each tree a little ways until he reached the last tree on the island, where he remained and listened in suspense. He soon heard a canoe on the river and someone calling the dogs. His predictions were about to come true, he thought. After making a fire, the man sent out his dogs. The man had a horrid-looking face, both behind and before [from the side], which the poor nephew could see by the firelight. Then the dogs began barking, having traced the tracks to the first tree. They made such a noise that the man thought they must have found game, and he went to the tree, but found nothing. So they went on to the next, and the next, still finding nothing, and this continued all night long. "There is no game here; my nephews have deceived me," the old man said angrily. And he left without checking the last tree.

After sunrise the poor young man came down from the tree and said, "I think I have escaped, for if those young fellows return, I will watch them and contrive to get their wings from them." He then concealed himself and patiently waited for them. He soon heard their voices saying, "Now we will have a good time." First they jumped around to warm themselves; then they said, "Let us all dive together." At this point the young man rushed out, and, taking all the wings, he put on one pair and flew away, calling out, "Uncle, now there is plenty of game for you." When they begged him to stop he replied, "You had no mercy on me; I only treat you the same." Then he flew on until he came home, where he found his old uncle, to whom he recounted the whole story. From that time on he remained peacefully at home with his good uncle.[27]

The half-human, half-diabolical Kingfisher in this Iroquois tale finds its opposite in the apotheosis of the bird in the mythology of the Bella Coola, a small tribe on the northwest coast of British Columbia. The latter tribe's complex pantheon associates the Belted Kingfisher with A't'mâk, one of nine divine brothers who live with their only sister in the "House of Myths" as watchers who monitor the winter religious ceremonial. The man representing A't'mâk in the ceremonial wears a Kingfisher mask shaped like the bird's head with wings out at the sides and tail feathers over its forehead.[28]

Measured against the fate of these native peoples, Kingfishers on both sides of the ocean are not faring too badly.

Afterword

MAIS TOI, animal,
plus je te regarde, ANIMAL,
plus je deviens HOMME
en Esprit.
PAUL VALÉRY[1]

IT IS a known fact that 133 bird species worldwide have become extinct since 1500, 103 of them since 1800. Their story tells us that rarity means jeopardy and that each species has its own threshold where danger begins. Take the Common Murre (*Uria aalge*), whose largest colonies number in the millions. In Brittany, where 320 breeding pairs are scattered over four colonies, they are thought to be so rare that *Uria aalge* is considered endangered in France. And yet the Common Murre is holding steady throughout the world. The global IUCN *Red List*[2] does not place the Common Murre in any of its four categories of threatened birds: critically endangered—extremely high risk of extinction in the wild; endangered—very high risk of extinction in the wild; vulnerable—high risk of extinction in the wild, and near threatened—likely to become endangered in the near future. From Albatross to Goatsucker, throughout the world there are 192 critically endangered species, 362 endangered, 669 vulnerable, and 838 near threatened.

As of 2010, twenty species in the large parrot family are already extinct, fifteen are critically endangered, and seventy-nine are more or less endangered. Brehm stated that a tropical forest without the parrot's bright feathers, chatter, whistles, and screams would be "a sorcerer's dead garden, a realm of silence, a wasteland."[3] In 1829, when he was born, there were a billion people on earth. Today there are more than six billion. More people lived between 1900 and 2000 than have existed from as long as mankind can remember up to 1900.[4] We have no Noah to stop the fragmentation of wildlife areas in this world flooded with our species. The story of Spix's Macaw teaches us how little the wise minority can counter stupidity and greed. Environmental protection is expensive. The poor of this world have more pressing concerns than the fact that most animals in their regions are endangered.

The humanizing of animals, so evident in Brehm's words, has long been taboo in science. References to obliging parrots, empathetic cranes, and wily vultures are supposed to be replaced by egoistic, altruistic, strategy-driven, and profit-maximizing genes—with people and animals functioning equally as their puppets. The more complicated the world, the simpler the hypotheses. If we have finally reduced the variety of cultures and the diversity of nature to values of natural selection, then we have found the theory of everything, and the human animal, in his hoary hubris, can use it to kid himself about the limits of his mind. As a reductionist, the most capable gene counter can only see organisms as "energy-acquiring systems . . . that live from generating a positive energy balance."[5] *Organisme* and *organisation* emerged in early eighteenth-century France to designate a whole that is composed of parts, "which all conspire together to produce a general result which we call 'life.'"[6] Once the *organisé* was taken from the natural sciences into sociology, it lost its sense of a harmonious interplay of parts that are committed to life. This new meaning carried over into biology when that discipline borrowed its metaphors from economics. Since then, "organization" and "organism" have been put into an inverted context. Originally introduced to characterize something alive and growing as opposed to something manufactured and mechanical, these terms now are associated with a process of mechanical control. Eibl-Eibesfeldt says about the genesis of "energy-acquiring systems":

> Their organization was constrained by natural selection; that is the very critical difference between organisms and physical systems. Competition forced many different kinds of adaptations concerning food acquisition and other battle lines of adaptation that were in principle always the same. So an organism has to adjust not only to a particular source of energy but also to the most diverse disruptive factors—enemies, climatic factors, or competitors. All this determines the commensurable costs for structuring and operations.[7]

Life: a school for maximizing benefit. Analogies can facilitate our understanding of scientific concepts. They lend the concepts certainty and the prestige of the fields the analogies are borrowed from. In the case of the theory of evolution, as refined by genetics, an analogy borrowed from biology was beneficial for *Homo economicus*. The law of the jungle made the law of the stock exchange now seem natural. Shareholders as well as apes were succumbing in all innocence to the constraints of almighty natural selection.

In the "Devil's gospel," as Darwin himself called the theory of evolution, pity for those weaker than oneself just maximizes genes.[8] If it does not, empathetic creatures will perish in the struggle for survival. Their genes will be weeded out—like those of all unfit creatures—on the "battle lines of adaptation" by natural selection. According to this logic, putting oneself on a level with alien species at the cost of one's own genes will lead to extinction. Nonetheless, sociobiologists bemoan the loss of biodiversity and urge man, as part of nature, to identify with other forms of life in order to save as many species as possible. It is unclear where the ability to put oneself in another creature's situation comes from, or how one can save something that is *before* him and *for* him at the same time. Since sociobiologists also erroneously believe that the ability to form aesthetic judgments or religious and moral convictions was derived from the mechanical process of unconscious natural selection, and since they regard the human mind as the executor of this process, they must invoke yet again the benefit of maximizing genes in order to justify their defense of biodiversity. Even Mother Earth is called upon to conjure up the "services" of species-rich ecosystems that range from garbage disposal and monitoring water quality and air purity to "the recreational value of pristine places" for tourists and to sources of medicines from flora and fauna—after all, everything with legs and feathers, all that grows and prospers, is supposed to serve the welfare of mankind, helped along by the pharmaceutical industry.[9] With that, the vicious circle is complete. The dilemma that the savior of species started from has caught up with him: "Our species appropriates between 20 and 40 percent of the solar energy captured in organic material by land plants. There is no way that we can draw upon the resources of the planet to this extent without drastically reducing the state of most other species."[10]

Supposedly, we don't have a code of ethics that enables us to "select" what is best for the short and long term, which is why sociobiologists are calling for a new ethics "uncoupled from other systems of belief."[11] Levi-Strauss, in *The View from Afar*, emphasizes the contradiction at the heart of sociobiological thinking:

[The sociobiological way of thinking] asserts, on the one hand, that all forms of the mind's activity are determined by inclusive fitness; and, on the other hand, that we can alter the fate of our species by deliberately choosing among the instinctive orientations that our biological past has transmitted to us. Yet the two things boil down to one: either the choices are themselves dictated by the demands of all-powerful inclusive fitness, and we are actually obeying it when we think we are choosing; or this possibility of choice is real, and nothing allows us to say that human destiny is ruled purely by our genetic heritage.[12]

Neuroscientists have appended a new chapter on unfree will to the Devil's gospel. "[Neural] connections determine us."[13] They claim that with every impulse, every thought secretly controlled by the brain, we've gained a new state of innocence that fairly cries out for a new ethics. We have no freedom of choice, so let's demonstrate that fact! You can't accuse scientists of not knowing what they're saying because that's what they say themselves. Whatever you shout into the forest of neurons will come back to you. In neurobiology's reductionistically attuned hemispheres of the brain, a negative answer to the question of free will is a given, just as it is in "insectolatry,"[14] which derives from the belief in a lack of freedom:

A colony of ants appears to be a closed system where everyone is dependent on everyone else. Individual ants do what they do because they are prompted to do it by all the others by means of many kinds of signals. Now we might imagine that it's exactly the same with humans except that the web of determinants is infinitely more complex. If we could observe ourselves from a higher perspective, we would find that we do this or that because these or those factors cause us to. These determinants naturally include our experience and our thoughts, but these all have a neuronal correlative. But down at our own level, we cannot view the multiplicity of parameters influencing us. Since we're not aware of this, it stands to reason that we impute intention to our acts, attribute intentionality to ourselves, and thus freedom.[15]

If we were completely determined, we would be just like the ants: we would live in a closed system with no questions or answers, insights or prospects. We would not know that we have no freedom, but we couldn't even know what freedom is. What's unintentionally funny about the example of the ants is the scrambling to justify it. The mind heaves itself up above itself in vain. From its higher standpoint, the mind can only discover what it already knows from its study of nerve cells: the pattern of stimulus and response. The mind clings so tightly to this standpoint that it is unaware of the inescapable bind it has got itself caught in. We can think that God is guiding us or the Devil is ordering us about and not violate the laws of logic. But the statement, "In all my doings, thinking and speaking, I am manipulated by my own brain," is a logical-semantic antinomy. If it were true, then the statement itself must have been dictated by neurons, free of any intention, so that its claim to truth would be invalid. But if the statement were false because it is not counted as a statement that is constrained by the brain, then it would invalidate itself (*sich aufheben*). The statement short circuits the train of thought in an alarming manner. Something that cannot be either true or false, but is said to be just the way it is, opens up the question of benefit or use, so dear to reductionists. Whether or not a neuroscientist's faith serves the gene maximization of its adherents is an unknown quantity in an evolution that is absolutely undeniable; but that belief contributes directly to an increase in nonsense, and that's the most innocuous thing you can say about it.

Our gateway to reality is language. We don't have any other way in. Precisely because we are denied direct access, we are not tied to one level and can choose from among different angles of vision. We are, as creatures of culture, located in another dimension compared with our fellow creatures that are perfectly at home in nature. However little leeway we have in making decisions, our dignity resides in the possibility of choice, which Pico della Mirandola has commented on so beautifully.

His *Oration on the Dignity of Man* is based on the assumption that God created man incomplete, unlike the perfect creatures in nature, so that man would have the freedom to shape himself into the form he might prefer. The striving for perfection is born out of deficiency. In the animal kingdom it is unnecessary. "[S]eeds pregnant with all possibilities, the germ of every form of life" have been placed in man. It is up to him to see which of these will be cultivated, which will wither. Man, from God's point of view a creature "neither of heaven nor of earth, neither mortal nor immortal,"[16] has been destined to mirror within nature the beauty of creation in a cosmos he himself has created; man is a creature that can also fail in his task because of his freedom, by bringing forth distorted images that can lead him and his kind astray.

If we presume to say that we are perfect creatures of nature and differ only by degree from the animals through our seemingly higher development, then we fail to recognize that the way we humans have been formed orients us toward language. The examples of the socio- and neurobiologists teach us that "the rejection of the anthropomorphic view of nature inevitably leads to man's becoming an anthropomorphism himself" and, as such, to man's fighting against himself.[17]

No matter if it's an ant or an amoeba, a bee or a bear, ferret or flounder, zebra or zebra finch, every species in the animal kingdom is the hub of its own particular environment into which it has adapted by means of the anatomical structure of its organism; each adaptation is carried out differently and always to perfection. "Only a superficial glance might make it seem as if all marine animals were living in a homogeneous world common to them all. Closer study teaches us that each of these thousands of different life forms has its own particular environment that is mutually conditioned by the animal's body plan."[18] The Barn Owl's world is tytocentric, the Kingfisher's alcedocentric, Spix's Macaw's cyanopsittacentric, and each animal focuses only on what is significant in its functional environment: food, prey, defense, and reproduction. Each animal receives external stimuli with its "receptor systems" (*Merknetze*), to which it reacts with effector systems (*Wirknetze*). The close interpenetration of internal and external worlds precludes the perception of things existing separately, precludes concrete objectification, distance, and permanence: "The forcible removal of a thing from its functional environment, where it normally belongs, abolishes that thing's identity in the world that animals perceive."[19] A wheel spider will ambush a fly or gnat caught in its wheel-shaped web in front of its tunnel-shaped dwelling; but if you put its prey into the tunnel, the spider will shrink away from its favorite delectable food as if from an enemy.

In closely meshed receptor and effecter networks, there is no place for a differentiating, symbolic language. Animals do not speak, even if they can learn individual words. In a human sense, they have nothing to say. Only by chance can language deceive the animal's ear within its sensory field, symbolically, as illustrated in this anecdote from the life of a tame Amazon parrot called Kuno.[20]

This beautiful green, yellow-browed bird slept in a cage that was left open during the day, and he imitated perfectly sentences he heard repeatedly. His favorite sentence, which was accompanied by his master ruffling the parrot's feathers, was, "Aren't you just the best little parrot?" His master would be greeted with "Hello, Carlo!" by his friends, and the parrot would say that, too, especially when he was alone and wanted Carlo to come.

One day, our good friend Carlo thought it would be nice to take his parrot for a walk in the woods so that Kuno could freely gad about in his natural element. The bird would of course not understand why his companion didn't rise into the air with him, and that was probably why he kept flying back to perch on Carlo's shoulder, to try to get him to fly. Said and done. The walk went as planned, until suddenly a Goshawk loomed in sight and caught the airborne parrot in its claws. Since Kuno was raised in the world of humans without his fellow birds, he had not learned the alarm calls that a Yellow-fronted Amazon emits when a raptor catches it; so he called out what he always did when he felt abandoned: "Hello, Carlo!" The parrot's unusual appearance must have already irritated the irascible Goshawk. But the parrot's "Hello, Carlo!"—something the hawk had never heard from his prey—startled him so much that he instinctively relaxed his talons, and Kuno was able to fly back to his good friend's shoulder. After that, Kuno never wanted to leave the house. He didn't find the whole business funny at all. He hadn't spoken to the Goshawk, of course. Alarm notes are intended for your own species. Since he considered humans to be his own kind, his piercing "Hello Carlo" was naturally meant for his master. He didn't have to think about what he was supposed to say or what to call to whom. The stimulus to give a warning call was his by nature. He learned how that call sounded by living at a wingless giant's side. While he lacked much that was necessary to exist in the wild, he had perfectly adapted, Yellow-fronted-Amazon-wise, to the cave of his humans. It would be quite a different matter if a human child were to be abandoned in the wild and raised by wolves. He would not regard his adoptive mother as one of his own kind but see himself as a wolf. As a human being, he would be without essence; as a wolf, he would be a naked, disadvantaged creature and no match for any wolf in the pack.

It was quite natural for the Amazon's owner to give his bird a human name. Lévi-Strauss took this custom as a sign that "everything conspires to make us think of the bird world as a metaphorical [counterpart to] human society."[21] His belief that there must be a similarly metaphorical relationship between the bird's brain and ours was to find further confirmation. In the case of birds, cognitive operations—the code for which is inscribed in the mammalian cortex that surrounds the mass of the brain held by the striatum—fall,

> as if through the effect of a topological transformation ... to the upper part of the striatum, which constitutes almost the whole mass of the brain, and which partly surrounds a rudimentary

cortex lodged in a furrow at the top... In so far as a metaphor always consists of referring to a total, implied semantic field by means of a complementary part of the whole, we can say, then, that in the field of possible cerebral organizations, the mammalian brain and the bird brain present a metaphorical image of each other.[22]

In the "Parliament of Birds," a late medieval genre of poems in stanza form, birds appear in the guise of birds of virtue and birds of vice. The wren (*Zaunkönig*, literally, "fence king"), as the smallest of the birds and the ruler of the kingdom, assembles his senior advisors to determine how to maintain the honor of the realm. Every piece of advice a virtuous species offers, such as the faithful dove or the generous eagle, is met with the bad counsel of a vicious species, like the merciless harrier or the greedy vulture. The oldest surviving "parliament of the birds" gives the last word, after forty-six stanzas of good and bad advice, to the Kingfisher. This wise old bird comes to the conclusion that the kingdom is missing out because the words of advice do not agree.[23]

"Birds, Birds, held by long affinity close to man's frontiers..." These are the lyrical words of Saint-John Perse, the pseudonym of Saint-Léger Léger.[24] "Man carries the weight of gravity like a millstone around his neck, the bird like a feather painted on his brow."[25] Freed from the force of gravity, birds seem to be intermediaries between heaven and earth, as the embodiments of flights of fancy. The creation story in Plato's *Timaeus* says that, of all the punishments the gods mete out to foolish men, the transformation into a bird is the most lenient. In the *Timaeus*, man, in a tradition harking back to Pythagoras's doctrine of the migration of souls, is a mutable creature. He changes shape according to gains and losses in reason or foolishness. The races of animals arose from foolish men through metamorphosis. Creatures of the waters, whether fish, mussel, sea urchin, or crab, originated from the most foolish men of all, whose totally corrupted souls God drove into the depths of the sea as punishment. God changed very foolish men into insensate land animals, condemned to crawl upon the ground, while he gave others the same number of feet that matched the degree of foolishness that pulled them down to the earth. Men who shunned the contemplation of the heavens were not quite that blinded but were bereft of all love of wisdom. Since they cleaved to the earth, they were turned into quadrupeds. But "the tribe of *birds* are derived by transformation, growing feathers in place of hair, from men who are harmless but light-minded—men, too, who, being students of the worlds above, suppose in their simplicity that the most solid proofs about such matters are obtained by the sense of sight."[26]

Saint Francis of Assisi, after being transformed from sinner to saint, called all animals his brothers. He understood a cicada's chirping as praise to the Lord, and since he harbored not the slightest doubt that the cicada's God was identical to his own, he addressed the insect as "Sister Cicada." His most intimate friendship seems to have been with the birds, his brothers the singers in Venice's marshes that praised their Creator for having given them feathers and wings; with the twittering swallows that were hushed at once and listened carefully when he spoke to them: "My sisters the swallows, it is now time for me to speak, for you have said enough and must be silent…"[27] Although he didn't know that even nocturnal birds like the owls are "sun worshippers," it befitted him as a friend of the birds that his most famous poem was *Il Cantico di Frate Sole*.

We may also consider, as a token of the intuitively understood relationships between bird and man, the comparison of an avian skeleton with a human one in Pierre Belon's *L'Histoire de la nature des oyseaux*. Linnaeus called botanical science *scientia amabilis*. He would probably not have objected in the slightest that ornithologists have appropriated this term for their own field because his love for birds was above all other animals. He expressed his admiration in the introduction of an essay on bird migration called *Migrationes avium*:

To the many good things with which the Almighty Creator has endowed man on this earth, for his use as well as for his joy, belongs the gift of birds. These have greatly appealed to my sense of beauty, and their appearance is so lovely that to my way of thinking there is nothing which contributes in a greater degree to the pleasure of the human race. What can compete with the humming-bird's dazzling beauty? What can surpass in splendour the peacock's colourful tail? These are, however, not nature's only works that are deserving of our attention: every bird, if we regard it with due care, presents a rich subject for admiration. We shall see them all gleam with the changing play of colours, which follow each other in the most ingenious succession and mutual rivalry—I doubt if the most skilful painter can so mix his colours as fully to reproduce nature in this respect.

If we regard birds' feathers and their peerless structure we shall clearly see that there is not a single feather which is not assembled by the great Artist with such skill that the whole of our short life span is insufficient to explore and rightly explain its mechanism and to muse upon His wisdom.[28]

Linnaeus's rival, Buffon, expressed his equally great admiration for feathered creatures in quite a different style. As enlightened as his depiction of birds' nature could possibly be for his time, his *Discours sur la nature des oiseaux* has lost none of its liveliness. Whatever in his treatise that would alienate neo-Darwinist ethologists bears witness to a different image of man than theirs. These behaviorists wouldn't dream of imputing human emotions to animals. They view it as their task to trace human behavior back to animal reactions. From this point of view, a smile appears to be a buffer against aggression, and the baring of teeth when laughing is "a threatening gesture turned into a welcoming ceremonial."[29]

"[F]ear of 'anthropomorphism' has never yet deterred anyone [from it]."[30] What ethologists call "ritualization" in animal behavior fulfills

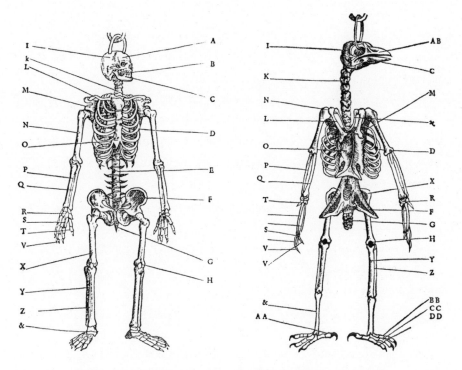

Pierre Belon, *Human and Avian Skeletons Compared*, 1555

the opposite purpose of ritual in the world of humans.[31] The long index to the textbook of comparative behavioral research cannot find a place for "love" anywhere from A to Z... Buffon, one of the soberest of men, regarded love as "Nature's soul! inexhaustible principle of existence! sovereign power that can do everything, and against which nothing prevails, by which everything acts, everything breathes and everything renews itself!" After celebrating love so enthusiastically as the "divine flame," he struggled with the question why love produces happiness for all living beings... and unhappiness for man.[32] The observation of animals led him to reflect on human customs; to do the opposite would never have entered his head. He could no more have empathized with Konrad Lorenz's explanation of laughter than ethologists could have empathized with his glorification of love in the avian realm. While ethologists are no longer able to grasp the difference between man and beast, Buffon overestimated it and deduced from man's mental powers that man was the superior being; but this permitted man to tyrannize all creatures—something Buffon altogether deplored. The prerogatives he conceded to people vis-à-vis animals he also endorsed for Europeans vis-à-vis people who were backward in their control over nature, like the aboriginals in Africa or America—whose appalling treatment at the hands of their white masters Buffon also deplored.

Whether we draw a non-humane border between man and beast or fail to recognize the difference between them, it all boils down to the same thing. As the most capable masters of nature and the most proficient gene maximizers, we believe we have the right to exploit every creature. Evolutionary biologists don't deny that all people deserve the same respect. But if they genuinely believed this, they would stop toying with the idea of improving the "human genotype" when "population thinking has gained further acceptance" and the horrendous deeds of twentieth-century eugenics have been forgotten.[33]

The approach to life that is reflected in aboriginal myths was as far removed from Buffon as it was from evolutionary biologists. This attitude toward life does not grant primacy to humans and confers the same religious dignity on life in all its forms.[34] In the modern world, where mankind is attuned to techniques for controlling nature, this view of life is considered obsolete; but everything argues that it constitutes the basic premise for preserving life on earth. And *la pensée sauvage* has still not been banished from poetry. At the limits of man, Saint-Léger Léger sees birds as congenial creatures, as strangers drawn nigh:

> But at dawn they come down to us, strangers descending: robed in those colours of dawn—between bitumen and hoarfrost—that are the very colours of the depths of man ... And from that dawn of freshness, as from a very pure aspersion, they have preserved for us something of the dream of creation.[35]

GEORGES LOUIS LECLERC,

COMTE DE BUFFON

Discourse on the Nature of Birds (1770)[1]

THE WORD Nature has in all languages two very different acceptations. It denotes either that Being, to the operations of which we usually ascribe the chain of effects that constitute the phænomena of the universe; or it signifies the aggregate of the qualities implanted in man, or in the various quadrupeds, and birds, &c. It is *active* nature that, stamping their peculiar characters, thus forms *passive* nature; whence are derived the *instincts* of animals, their *habits*, and their *faculties*. We have in a former work treated of the nature of Man and the Quadrupeds; that of Birds now demands our attention: and though the subject is, in many respects, more obscure, we shall endeavour to select the discriminating features, and to place them in the proper point of view.

Perception, or rather the faculty of feeling; instinct, which results from it; and talent, which consists in the habitual exercise of the natural powers; are widely distinguished in different beings. These intimate qualities depend upon organization in general, and especially upon that of the senses: they are not only proportioned to the degree of the perfection of these; they have also a relation to the order of superiority that is established. In man, for instance, the sense of touch is more exquisite than in all other animals:[2] in these, on the contrary, smell is more perfect than in man: for touch is the foundation of knowledge, and smell is only the source of perception. But, as few persons distinguish nicely the shades that discriminate between ideas and sensations, knowledge and perception, reason and instinct, we shall set aside what are termed *ratiocination, discernment,* and *judgment;* and we shall only consider the different combinations of simple perception, and endeavour to investigate the causes of that diversity of instinct, which, though infinitely varied in the immense number of species, seems more constant, more uniform, and more regular, and less subject to caprice and error, than reason in the single species which boasts the possession of it.

In comparing the senses, which are the primary powers that readily excite and impel the instinct in all animals, we find that of sight to be more extended, more acute, more accurate, and more distinct in the birds in general, than in the quadrupeds: I say in general, for there are some birds, such as the owls, that have less clear vision than the quadrupeds; but this, in fact, results from the excessive sensibility of the eye, which, though it cannot support the glare of noon-day, distinguishes nicely objects in the glimmering of the evening. In all birds the organ of sight is furnished with two membranes, an external and internal, additional to those which occur in man: the former [*membrana nictitans*], or external membrane, is placed in the large angle of the eye, and is a second and more transparent eye-lid, whose motions too are directed at pleasure, and whose use is to clear and polish the cornea: it serves also to temper the excess of light, and consequently to adjust the quantity admitted, to the extreme delicacy of the organ: the other is situated at the bottom of the eye, and appears to be an expansion of the optic nerve, which, receiving more immediately the impressions of the rays, must be much more sensible than in other animals; and hence the sight is in birds vastly more perfect, and embraces a wider range. A sparrow-hawk, while he hovers in the air, espies a lark sitting on a clod, though at twenty times the distance at which a man or dog could perceive it. A kite which soars to so amazing a height as totally to vanish from our sight, yet distinguishes the small lizards, field-mice, birds, &c. and from this lofty station he selects what he destines to be victims of his rapine. But this prodigious extent of vision is accompanied likewise with an equal accuracy and clearness; for the eye can dilate or contract, can be shaded or uncovered, depressed or made protuberant, and thus it will readily assume the precise form suited to the quantity of light and the distance of the object.

Sight has a reference also to motion and space; and, if birds trace the most rapid course, we might expect them to possess in a superior degree that sense which is proper to guide and direct their flight. If Nature, while she endowed them with great agility and vast muscular strength, had formed them short-sighted, their latent powers would have availed them nothing; and the danger of dashing against every intervening obstacle would have repressed or extinguished their ardour. Indeed, we may consider the celerity with which an animal moves, as the just indication of the perfection of its vision. A bird, for instance, that shoots swiftly through the air, must undoubtedly see better than one which slowly describes a waving tract. Among the quadrupeds too, the *sloths* have their eyes enveloped, and their sight is limited.

The idea of motion, and all the other ideas which accompany or flow from it, such as those of relative velocities, of the extent of country, of the proportional height of eminences, and of the various inequalities that prevail on the surface, are, therefore, more precise in birds, and occupy a larger share of their conceptions than in quadrupeds. Nature would seem to have pointed out this superiority of vision by the more conspicuous and more elaborate structure of its organ; for in birds the eye is larger in proportion to the bulk of the head than in quadrupeds; it is also more delicate and more finely fashioned, and the impressions which it receives must excite more vivid ideas.

Another cause of the difference between the instincts of birds and of quadrupeds, is the nature of the element in which they live. The birds know better than man, perhaps, all the degrees of resistance of the air, its temperature at different heights, its relative density, &c. They foresee more than us, they indicate better than our barometers or thermometers, the changes which happen in that voluble fluid. Often have they struggled against the violence of the wind, and oftener have they borrowed its aid. The eagle, soaring above the clouds, can quickly escape from the scene of the storm to the region of calm, and there enjoy a serene sky and a bright sun, while the other animals below are involved in darkness, and exposed to all the fury of the tempest. In twenty-four hours it can change its climate, and sailing over the different countries, it will form a picture which exceeds the powers of our imagination. Our bird's-eye views, of which the accurate execution is so tedious and so difficult, give very imperfect notions of the relative inequality of the surfaces which they represent. But birds can chuse the proper stations, can successively traverse the field in all directions, and with one glance comprehend the whole. The quadruped knows only the spot where it feeds; its valley, its mountain, or its plain: it has no conception of the expanse of surface, no idea of immense distances, and no desire to push forward its excursions. Hence remote journies and migrations are as rare among the quadrupeds as they are frequent among the birds.[3] It is this desire, founded on their acquaintance with foreign countries, on the consciousness of their expeditious course, and on their foresight of the changes that will happen in the atmosphere and of the revolution of seasons, that prompt them to retire together, and by common consent. When their food begins to grow scarce, when, as the cold or the heat incommodes them, they resolve on their retreat, the parents collect their young,[4] and the different families assemble and communicate their views to the unexperienced; and the whole body, strengthened by their numbers, and actuated by the same common motives, wing their journey to some distant land.

This propensity to migration, which recurs every spring and autumn, is a sort of violent longing, which, even in captive birds, bursts out in symptoms of restless and uneasy sensations. We shall, at the article of the Quail, give a detail of observations on this subject; from which it will appear, that this propensity is one of their most powerful instincts; and that, though they usually remain tranquil in their prison, they make every exertion at those periods to regain their liberty, and join their companions.—But the circumstances which attend migration vary in different birds; and, before we enter into the full discussion which that subject merits, we shall pursue our investigation of the causes that form and modify their instincts.

Man is eminently superior to all the animals in the sense of touch, perhaps too in that of taste; but he is inferior to most of them in the other three senses. When we compare the animals with each other, we soon perceive that smell in general is more acute among the quadrupeds than among the birds: for though we speak of the scent of the crow, of the vulture, &c. it undoubtedly obtains in a much lower degree; and we might be convinced of this by merely examining the structure of the organ. In most of the winged tribes, the external nostrils are wanting,[5] and the effluvia, which excite the sensation, have access only to the duct leading from the palate: and even in those where the organ is disclosed, the nerves, which take their origin from it, are far from being so numerous, so large, or so expanded, as in the quadruped. We may therefore regard touch in man, smell in the quadruped, sight in the bird, as the three most perfect senses, and which influence the general character.[6]

Next to sight, the most perfect of the senses in birds is hearing, which is even superior to that of the quadrupeds. We perceive with what facility they retain and repeat tones, successions of notes, and even discourse; we delight to listen to their unwearied songs, to the incessant warbling of their happy loves. Their ear and throat are more ductile and more powerful than in other animals. Most of the quadrupeds are habitually silent; and their voice, which is seldom heard, is almost always harsh and disagreeable. In birds it is sweet, pleasant, and melodious. There are some species, indeed, in which the notes seem unsupportable, especially if compared with those of others; but these are few in number, and comprehend the large kinds, which Nature, bestowing on them hoarse loud cries, suited to their bulk, would incline to treat like quadrupeds. A pea-cock, which is not the hundredth part of the size of an ox, may be heard farther; the nightingale could fill a wider space with its music than the human voice: this prodigious extent and the great powers of their organs of sound, depend entirely on the structure; but that their song should be continued and supported, results solely from their internal emotions. These two circumstances ought to be considered separately.

The pectoral muscles are more fleshy and much stronger in birds than in man or the quadrupeds, and their action is immensely greater. Their wings are broad and light, composed of thin hollow bones, and con-nected by powerful tendons. The ease with which birds fly, the celerity of their course, and even their power of directing it upwards or down-wards, depend on the proportion of the impelling surface to the mass of the body. When they are ponderous, and the wings and tail at the same time short, like the bustard, the cassowary, or the ostrich, they can hardly rise from the ground.

The windpipe is wider and stronger in birds than in quadrupeds, and usually terminates below in a large cavity that augments the sound. The lungs too have greater extent, and send off many appendices which form air-bags, that at once assist the motion, by rendering the body specifically lighter, and give additional force to the voice. A little production of the cartilage of the *trachea* in the howling baboon, which is a quadruped of a middle size only, and of the ordinary structure, has enabled it to scream almost without intermission, and so loud, as to be heard at more than a league's distance: but in birds, the formation of the thorax, of the lungs, and of all the organs connected with these, seems expressly calculated to give force and duration to their utterance; and the effect must be propor-tionally greater.

There is another circumstance which evinces that birds have a prodi-gious power of voice: the cries of many species are uttered in the higher regions of the atmosphere, where the rarity of the medium must con-sequently weaken the effect. That the rarefaction of the air diminishes sounds is well ascertained from pneumatical experiments; and I can add, from my own observation, that, even in the open air, a sensible difference in this respect may be perceived. I have often spent whole days in the forests, where I was obliged to listen closely to the distant cries of the dogs, or shouts of the hunters; I uniformly found that the same noises were much less audible during the heat of the day, between ten and four o'clock, than in the evening, and particularly in the night, whose stillness would make hardly any alteration, since in these sequestered scenes there is nothing to disturb the harmony but the flight buzz of insects[7] and the chirping of some birds. I have observed a similar difference between the frosty days in winter and the heats of summer. This can be imputed only to the variation in the density of the air. Indeed, the difference seems to be so great, that I have often been unable to distinguish in mid-day, at the distance of six hundred paces, the same voice which I could, at six o'clock in the morning or evening, hear at that of twelve or fifteen hundred paces.—A bird may rise at least to the height of seventeen thousand feet, for it is there just visible. A flock of several hundred storks, geese, or ducks, must mount still higher, since, notwithstanding the space which they occupy, they soar almost out of sight. If the cry of birds therefore may be heard from an altitude of above a league, we may reckon it at least four times as powerful as that of men or quadrupeds, which is not audible at more than half a league's distance on the surface. But this estimation is even too low: for, beside the dissipation of force to be attributed to the cause already assigned, the sound is propagated in the higher regions as from a centre in all directions, and only a part of it reaches the ground; but, when made at the surface, the aerial waves are reflected as they roll along, and the lateral and vertical effect is augmented. It is hence that a person on the top of a tower hears one better at the bottom, than the person below hears from above.

Sweetness of voice and melody of song are qualities which in birds are partly natural, partly acquired. Their great facility in catching and repeating sounds enables them not only to borrow from each other, but often to copy the inflexions and tones of the human voice, and of our musical instruments. Is it not singular, that in all populous and civilized countries, most of the birds chant delightful airs, while, in the extensive deserts of Africa and America, inhabited by roving savages, the winged tribes utter only harsh and discordant cries, and but a few species have any claim to melody?[8] Must this difference be imputed to the difference of climate alone? The extremes of cold and heat operate indeed great changes on the nature of animals, and often form externally permanent characters and vivid colours. The quadrupeds of which the garb is variegated, spotted, or striped, such as the panthers, the leopards, the zebras, and the civets, are all natives of the hottest climates. All the birds of the tropical regions sparkle with the most glowing tints, while those of the temperate countries are stained with lighter and softer shades. Of the three hundred species that may be reckoned belonging to our climates, the peacock, the common cock, the golden oriole, the king-fisher, and the goldfinch, only can be celebrated for the variety of their colours; but Nature would seem to have exhausted all the rich hues of the universe on the plumage of the birds of America, of Africa, and of India. These quadrupeds, clothed in the most splendid robes, these bird attired in the richest plumage, utter at the same time hoarse, grating, or even terrible cries. Climate has no doubt a principal share in this phænomenon; but does not the influence of man contribute also to the effect? In all the domesticated animals, the colours never heighten, but grow softer and fainter: many examples occur among the quadrupeds; and cocks and pigeons are still more variegated than dogs or horses. The real alteration which the human powers have produced on nature, exceeds our fondest imagination: the whole face of the globe is changed; the milder animals are tamed and subdued, and the more ferocious are repressed and extirpated. They imitate our manners; they adopt our sentiments; and, under our tuition, their faculties expand. In the state of nature, the dog has the same qualities and dispositions, though in an inferior degree, with the tiger, the leopard, or the lion; for the character of the carnivorous tribe results solely from the acuteness of their smell and taste: but education has mollified his original ferocity, improved his sagacity, and rendered him the companion and associate of man.

Our influence is smaller on the birds than on the quadrupeds, because their nature is more different from our own, and because they are less submissive and less susceptible of attachment. Those we call *domestic*, are only prisoners, which, but for propagating, are useless during their lives; they are victims, multiplied without trouble, and sacrificed without regret. As their instincts are totally unrelated to our own, we find it impossible to instil our sentiments; and their education is merely mechanical. A bird, whose ear is delicate, and whose voice is flexible, listens to discourse, and soon learns to repeat the words, but without feeling their force. Some have indeed been taught to hunt and fetch game; some have been trained to fondle their instructor: but these sentiments are infinitely below what we communicate so readily to the quadrupeds. What comparison between the attachment of a dog, and the familiarity of a canary bird; between the understanding of an elephant, and the sagacity of an ostrich [which seems to be the worthiest and cleverest of birds]⁹?

The natural tones of birds, setting aside those derived from education, express the various modifications of passion; they change even according to the different times or circumstances. The females are much more silent than the males; they have cries of pain or fear, murmurs of inquietude or solicitude, especially for their young; but song is generally withheld from them. In the male it springs from sweet emotion, from tender desire; the canary in his cage, the greenfinch in the fields, the oriole in the woods, chant their loves with a sonorous voice, and their mates reply in feeble notes of consent. The nightingale, when he first arrives in the spring, is silent; he begins in faultering unfrequent airs: it is not until the dam sits on her eggs, that he pours out the warm melody of his heart: then he relieves and soothes her tedious incubation; then he redoubles his caresses, and warbles more pathetically his amorous tale.[10] And what proves that love is among birds the real source of their music is, that, after the breeding season is over, it either ceases entirely, or loses its sweetness.

This melody, which is each year renewed, and which lasts only two or three months during the season of love, and changes into harsh low notes on the subsidence of that passion, indicates a physical relation between the organs of generation and those of voice, which is most conspicuous in birds. It is well known that the articulation is never confirmed in the human species before the age of puberty; and that the bellowing of quadrupeds becomes tremendous when they are actuated by their fiery lusts. The repletion of the spermatic vessels irritates the parts of generation, and by sympathy affects the throat. Hence the growth of the beard, the forming of the voice, and the extension of the genital organ in the male; the swell of the breasts, and the expansion of the glandulous bodies in the female. In birds the changes are more considerable; not only are these parts stimulated or altered; after being in appearance entirely destroyed, they are even renovated by the operation of the same causes. The testicles, which in a man and most of the quadrupeds remain nearly the same at all times, contract and waste almost entirely away in birds after the breeding season is over, and on its return they expand to a size that even appears disproportioned. It would be curious to discover if there is not some new production in the organs of the voice, corresponding to this swell in the parts of generation.

Man seems even to have given a direction to love, that appetite which Nature has the most deeply implanted in the animal frame. The domestic quadrupeds and birds are almost constantly in season, while those which roam in perfect freedom are only at certain stated times stimulated by the ardour of passion. The cock, the pigeon, and the duck, have, equally with the horse, the ram, and the dog, undergone this important change of constitution.

But the birds excel the other animals in the powers of generation, and in their aptitude for motion. Many species scarcely rest a single moment, and the rapacious tribes pursue their prey without halting or turning aside, while the quadrupeds need to be frequently recruited.—To give some idea of the rapidity and continuance of the flight of birds, let us compare it with the celerity of the fleetest land-animals. The stag, the rein-deer, and the elk, can travel forty leagues a-day; the rein-deer can draw its sledge at the rate of thirty leagues for several days. The camel can perform a journey of three hundred leagues in eight days. The choicest race-horse can run a league in six or seven minutes; but he soon slackens his career, and could not long support such an exertion. I have elsewhere mentioned the instance of an Englishman who rode sixty-two leagues in eleven hours and thirty-two minutes, changing horses twenty-one times: so that the best horse could not travel more than four leagues in an hour, or thirty leagues in a day. But the motion of birds is vastly swifter: an eagle, whose diameter exceeds four feet, rises out of sight in less than three minutes, and therefore must fly more that 3,500 yards in one minute, or twenty leagues in an hour. At this rate, a bird would easily perform a journey of two hundred leagues in a day, since ten hours would be sufficient, which would allow frequent halts, and the whole night for repose. Our swallows, and other migratory birds, might therefore reach the equator in seven or eight days. [Michel] Adanson saw on the coast of Senegal swallows that had arrived on the ninth of October; that is, eight or nine days after their departure from Europe. Pietro della Valle says, that in Persia the messenger-pigeon travels as far in a single day as a man can go a-foot in six days. It is a well-known story, that a falcon of Henry ii. which flew after a little bustard at Fontainbleau, was caught next morning at Malta, and recognized by the ring which it wore. A Canary falcon, sent to the duke of Lerma, returned in sixteen hours from Andalusia to the island of Teneriffe, a distance of two hundred and fifty leagues. Sir Hans Sloane assures us, that at Barbadoes the gulls make excursions in flocks to the distance of more than two hundred miles, and return the same day. Taking all these facts together, I think we may conclude that a bird of vigorous wing could every day pass through four or five times more space than the fleetest quadruped.

Every thing conspires to the rapidity of a bird's motion: first, the feathers are very light, have a broad surface, and their shafts are hollow: secondly, the wings are convex above and concave below; they are firm and wide spread, and the muscles which act upon them are powerful: thirdly, the body is proportionally light, for the flat bones are thinner than in the quadrupeds, and hollow bones have much larger cavities. "The skeleton of the pelican," say the anatomists of the Academy, "is extremely light, not weighing more than twenty-three ounces, though it is of considerable bulk." This quality diminishes the specific gravity of birds.

Another consequence which seems to result from the texture of the bones, is the longevity of birds. In man and the quadrupeds, the period of life seems to be in general regulated by the time requited to attain the full growth: but in birds it follows different proportions; their progress is rapid to maturity; some run as soon as they quit the shell, and fly shortly afterward: a cock can copulate when only four months old, and yet does not acquire his full size in less that a year.[11] Land animals generally live six or seven times as long as they take to reach the age of puberty; but in birds the proportion is ten times greater, for I have seen linnets fourteen or fifteen years old, cocks twenty, and parrots above thirty, and they would probably go beyond these limits. This difference I should attribute to the soft porous quality of the bones; for the general ossification and rigidity of the system to which animals perpetually tend, determine the boundary of life; that will therefore be prolonged, if the parts want solidity and consistence. It is thus that women arrive oftener at old age than men; that birds live longer than quadrupeds, and that fishes live longer than birds.

But a more particular inquiry will evince that uniformity of plan which prevails through nature. The birds, as well as the quadrupeds, are carnivorous, or granivorous. In the former class, the stomach and intestines are proportionally small; but those of the latter have a craw additional, corresponding to the false belly in ruminating animals, and the capacity of the ventricle compensates for the unsubstantial quality of their destined food. The granivorous birds have also two *cæca,* and a very strong muscular stomach, which serves to triturate the hard substances which they swallow.

The dispositions and habits of animals depend greatly on their original appetites. We may therefore compare the eagle, noble and generous, to the lion; the vulture, cruel and insatiable, to the tiger; the kite, the buzzard, the crow, which only prowl among carrion and garbage, to the hyænas, the wolves, and jackals. The falcons, the sparrow-hawks, the gos-hawks, and the other birds trained for sport, are analogous to the dogs, the foxes, the ounces [i.e., snow leopards, *Panthera uncia*], and the lynxes; the owls, which prey in the night, represent the cats; the herons, and the cormorants, which live upon fish, correspond to the beavers and otters; and, in their mode of subsistence, the woodpeckers resemble the ant-eaters. The common cock, the peacock, the turkey, and all the birds furnished with a craw, bear a relation to the ox, the sheep, the goat, and other ruminating animals.[12] With regard to the article of food, birds have a more ample latitude than quadrupeds; flesh, fish, the amphibious tribes, reptiles, insects,[13] fruits, grain, seeds, roots, herbs; in a word, whatever lives or vegetates. Nor are they very nice in their choice, but often catch indifferently at what they can most easily obtain. The sense of taste is much less acute in birds that in quadrupeds; for, if we except such as are carnivorous, their tongue and palate are in general hard, and almost cartilaginous. Smell can alone direct them,[14] and this they possess in an inferior degree. The greater number swallow without tasting, and mastication, which constitutes the chief pleasure in eating, is entirely wanting to them. Hence, on all these accounts, they are so little attentive to the selection of their food, that they often poison themselves.

The attempt is impossible therefore to distinguish the winged tribes according to the nature of their aliments. The more constant and determined appetites of quadrupeds might countenance such a division; but in birds, where the taste is so irregular, it would be entirely nugatory. We see hens, turkies, and other fowls which are called granivorous, eat worms, insects, and bits of flesh with greater avidity than grain. The nightingale, which lives on insects, may be fed with minced meat; the owls, which are naturally carnivorous, often when other prey fails, catch night-flies in the dark;[15] nor is their hooked bill, as those who deal in final causes maintain, any certain proof that they have a decided propensity for flesh, since parrots and many other birds which seem to prefer grain have also a hooked bill. The more voracious kinds devour fish, toads, and reptiles, when they cannot obtain flesh. Almost all the birds which appear to feed upon grain, were reared by their parents with insects. The arrangement derived from the nature of the food is thus totally destitute of foundation. No one character is sufficient: it requires the combination of many.

Since birds cannot chew, and the mandibles which represent the jaws are unprovided with teeth, the grains are swallowed whole, or only half-bruised. But the powerful action of the stomach serves them instead of mastication; and the small pebbles, which assist in trituration, may be conceived to perform the office of teeth.

As Nature has invested the quadrupeds which haunt marshes, or inhabit cold countries, with a double fur, and with thick close hair; so has she clothed the aquatic birds, and those which live in the northern tracts, with abundance of plumage, and a fine down; insomuch that, from this circumstance alone, we may judge of their proper element, or of their natal region. In all climates, the birds which dwell in the water are nearly equally feathered, and have under the tail large glands, containing an oily substance for anointing their plumes, which, together with their thickness, prevents the moisture from insinuating. These glands are much smaller in the land-birds, or totally wanting.

Birds that are almost naked, such as the ostrich, the cassowary, and the dodo, occur only in the warm climates. All those which inhabit cold countries are well clothed with plumage. And for the same reason, those which soar into the higher regions of the atmosphere require a think covering, that they may encounter the chilness which there prevails. If we pluck the feathers from the breast of an eagle, he will no longer rise out of our sight.

The greater number of birds cast their feathers every year, and appear to suffer much more from it than the quadrupeds do from a similar change. The best fed hen ceases at that time to lay. The organic molecules seem then to be entirely spent on the growth of the new feathers. The season of moulting is generally the end of summer or autumn, and their feathers are not completely restored till the beginning of spring, when the mildness of the air, and the superabundance of nutrition, urge them to love. Then all the plants shoot up, the insects awaken from their long slumber, and the earth swarms with animation. This ample provision fosters their ardent passions, and offers abundant subsistence to the fruits of their embrace.

We might deem it as essential to the bird to fly, as it is to the fish to swim, or to the quadruped to walk; yet in all these tribes there are exceptions to the general property. Among quadrupeds the rufous, red and common bats, can only fly; the seals, the sea-horses, and sea-cows, can only swim; and the beavers and otters walk with more difficulty than swim: and, lastly, there are others, such as the sloth, which can hardly drag along their bodies. In the same manner, we find among birds the ostrich, the cassowary, the dodo, the touyou, &c. which are incapable of flying, and are obliged to walk; others, such as the penguins, the sea-parrots, &c. which fly and swim, but never walk: and others, in fine, which, like the bird of paradise, can neither walk nor swim, but are perpetually on the wing. It appears, however, that water is, on the whole, more suited to the nature of birds than to that of quadrupeds: for, if we except a few species, all the land animals shun that element, and never swim, unless they are urged by their fears or wants. Of the birds, on the contrary, a large tribe constantly dwell on the waters, and never go on shore, but for particular purposes, such as to deposite their eggs, &c. And what proves this position, there are only three or four quadrupeds which have their toes connected by webs; whereas we may reckon above three hundred birds which are furnished with such membranes. The lightness of their feathers and of their bones, and even the shape of their body, con-tribute greatly to the facility with which they swim, and their feet serve as oars to impel them along. Accordingly, certain birds discover an early propensity to the water; the ducklings sail on the surface of the pool long before they can use their wings.

In quadrupeds, especially those which have their feet terminated by hard hoofs or nails, the palate seems to be the principal seat of touch as well as of taste. Birds, on the other hand, oftener feel bodies with their toes; but the inside of these is covered with a callous skin, and their tongue and mouth are almost cartilaginous: so that, on both accounts, their sensations must be blunt.[16]

Such then is the order of the senses which Nature has established in the different beings. In man, touch is the first, or the most perfect; taste the second; sight the third; hearing the fourth; and smell the fifth and last. In quadrupeds, smell is the first; taste the second, or rather these two senses form only one; sight the third; hearing the fourth; and touch the last. In birds, sight is the first; hearing the second; touch the third; and taste and smell the last. The predominating sensations will also fol-low the same order: man will be most affected by touch; the quadrupeds by smell; and the birds by sight. These will likewise give a cast to the gen-eral character, since certain motives of action will acquire peculiar force, and gain the ascendency. Thus, man will be more thoughtful and pro-found, as the sense of touch would appear to be more calm and intimate; the quadrupeds will have more vehement appetites; and the birds will have emotions as extensive and volatile as is the glance of sight.

But there is a sixth sense, which, though it intermits, seems, while it acts, to control all the others, and excites the most powerful emotions, and awakens the most ardent affections:—it is love. In quadrupeds, that appetite produces violent effects; they burn with maddening desire; they seek the female with savage ardor; and they embrace with furious extasy. In birds it is a softer, more tender, and more endearing passion; and, if we except those which are degraded by domestication, and a few other species, conjugal fidelity and parental affection are among them alike conspicuous.[17] The pair unite their labours in preparing for the accommodation of their expected progeny; and, during the time of incubation, their participation of the same cares and solicitudes continually augments their mutual attachment. After the eggs are hatched, a new source of pleasure opens to them, which further strengthens the ties of affection; and the tender charge of rearing the infant brood requires the joint attention of both parents. The warmth of love is thus succeeded by calm and steady attachment, which by degrees extends, without suffering any diminution, to the rising branches of the family.

The quadrupeds are impelled by unbridled lust, which never softens into generous friendship. The male abandons the female as soon as the cravings of his appetite are cloyed; he retires to recruit his strength, or hastens to the embraces of another. The education of the young is devolved entirely on the female; and as they grow slowly, and require her immediate protection, the maternal tenderness is ripened into a strong and durable attachment. In many species the mother leads two or three litters at one time. There are some quadrupeds, however, in which the male and female associate together; such are the wolves and foxes: and the fallow-deer have been regarded as the patterns of conjugal fidelity. There are also some species of birds where the cock separates after satisfying his passion—but such instances are rare, and do not affect the general law of nature.

That the pairing of birds is founded on the need of their mutual labours to the support of the young, appears clearly from the case of the domestic fowls. The male ranges at will among a seraglio of submissive concubines; the season of love has hardly any bounds; the hatches are frequent and tedious; the eggs are often removed; and the female never seeks to breed, until her prolific powers are deadened, and almost exhausted: besides, they bestow little care in making their nest, they are abundantly supplied with provisions, and by the assistance of man they are freed from all those toils and hardships and solicitudes which other birds feel and share in common. They contract the vices of luxury and opulence, *indolence* and *debauchery*.

The easy comfortable condition of the domestic fowls, and their generous food, mightily invigorate the powers of generation. A cock can tread twelve or fifteen hens, and each embrace continues its influence for three weeks; so that he may each day be the father of three hundred chickens. A good hen lays a hundred eggs between the spring and autumn; but in the savage state she has only eighteen or twenty, and that only during a single season. The other birds indeed repeat oftener their incubations, but they lay fewer eggs. The pigeons, the turtles, &c. have only two; the great birds of prey three or four; and most other birds five or six.

Want, anxiety, and hard labour, check in all animals the multiplication of the species. This is particularly the case with birds; they breed in proportion as they are well fed, and afforded ease and comfort. In the state of nature, they seem even to husband their prolific powers, and to limit the number of their progeny to the penury of their circumstances. A bird lays five eggs, perhaps, and devotes her whole attention during the rest of the season to the incubation and education of the young. But if the nest be destroyed, she soon builds another, and lays three or four eggs more; and if this be again plundered, she will construct a third, and lay still two or three eggs. During the first hatch, therefore, those internal emotions of love which occasion the growth and exclusion of the eggs, are repressed. She thus sacrifices passion to duty, amorous desire to parental attachment. But when her fond hopes are disappointed, she soon ceases to grieve; the procreative faculties, which were suspended, not extinguished, again resume their influence, and enable her in some measure to repair her loss.

As love is a purer passion in birds than in quadrupeds, its mode of gratification is also simpler. Coition is performed among them only in one way, while many other animals embrace in various postures: only in some species, as in that of the common cock, the female squats; and in others, such as the sparrows, she continues to stand erect. In all of them the act is transitory, and is still shorter in those which in their ordinary attitude wait the approach of the male, than in those which cower to receive him. The external form, and the internal structure of the organs of generation are very different from what obtains in quadrupeds. The size, the position, the number, the action and motion of these parts even vary much in the several species of birds. In some there appears to be a real penetration; in others, a vigorous compression, or slight touch. But we shall consider the details in the course of the work.

To concentrate the different principles established in this discourse: that the *sensorium* of birds contains chiefly the images derived from the sense of sight; and these, though superficial, are very extensive, and, for the most part, relate to motion, to distance, and to space: that comprehending a whole province within the limits of their horizon, they may be said to carry in their brain a geographical chart of the places which they view: that their facility in traversing wide territories is one of the causes which prompt their frequent excursions and migrations:[18] that their ear being delicate, they are alarmed by sudden noises, but may be soothed by soft sounds, and allured by calls: that their organs of voice being exceedingly powerful and soft, they naturally vent their feelings in loud resounding strains: that, as they have more signs and inflexions, they can, better than the quadrupeds, express their meaning: that easily receiving, and long retaining the impressions of sounds, the organ delights in repeating them; but that its imitations are entirely mechanical, and have no relation to their conceptions: that their sense of touch being obtuse, they have only imperfect ideas of bodies: that they receive their information of distant objects from sight, not from smell: that as their taste is indiscriminating, they are more prone to voracity than sensuality: that, from the nature of the element which they inhabit, they are independent of man, and retain their natural habits; that, for this reason, most of them are attached to the society of their fellows, and eagerly convene: that, being obliged to unite their exertions in building a nest, and in providing for their offspring, the pair contract an affection for each other, which continues to grow, and then extends to the tender brood: that this friendship restrains the violent passions, and even tempers love, and begets chastity, and purity of manners, and gentleness of disposition: that, though their power of fruition is greater than in other animals, they confine its exercise within moderate bounds, and ever subject their pleasures to their duties: and, finally, that these sprightly beings, which Nature would seem to have produced in her gay moments, may be regarded as a serious and decent race, which exhibit excellent lessons and laudable examples of morality.

Endnotes

CHAPTER 1

1 "So sterben dahin die Geschlechter der
Menschen. Es verhallt die rühmliche
Kunde der Völker. Doch wenn jede
Blüte des Geistes welkt, wenn im Sturm
der Zeiten die Werke schaffender Kunst
zerstieben, so entsprießt ewig neues
Leben aus dem Schoße der Erde. Rast-
los entfaltet ihre Knospen die zeugende
Natur: unbekümmert, ob der frevelnde
Mensch (ein nie versöhntes Geschlecht)
die reifende Frucht zertritt." Alexander
von Humboldt, "Über die Wasserfälle
des Orinoco" (1849).

2 John James Audubon, *Ornithological
Biography* (Philadelphia: Dobson,
1831–39), 1:321.

3 Ibid.

4 Rev. 19:17.

5 Audubon, *Ornithological Biography*,
1:323–24.

6 A sole missionary took the Natives'
warning to heart and pleaded, without
success, on the pigeons' behalf; see David
Quammen, *The Song of the Dodo* (New
York: Scribner, 1996), 311.

7 See Christopher Cokinos, *Hope Is the
Thing with Feathers: A Personal Chronicle
of Vanished Birds* (New York: Tarcher/
Putnam, 2000), 228–44.

8 See Claus Nissen, *Die illustrierten
Vogelbücher* (Stuttgart: Hiersemann,
1953), 60.

9 Audubon, *Ornithological Biography*,
1:326.

10 [From "Audubon," a long poem by
Ferdinand Freiligrath (1810–76).
It was written in the same meter as
Longfellow's famous "Song of
Hiawatha" and published in 1848.
Its opening line addresses Audubon,
who died in 1851, as a "Mann der
Wälder, der Savannen!" Line seven
reads, appropriately enough, "[You]
saw the Passenger Pigeons' jour-
neys" ("Sahst der Wandertauben
Reisen").—TRANS.]

CHAPTER 2

1 Audubon, *Ornithological Biography*,
1:136.

2 Ibid.

3 Alexander Wilson, *Wilson's American Ornithology* (Boston: Otis Broaders & Co., 1840), n.p.

4 See Cokinos, *Hope Is the Thing with Feathers*, 18.

5 Wilson, *Wilson's American Ornithology*, n.p.

6 Eugène Rey, quoted in Alfred Edmund Brehm, *Brehms Tierleben: Vögel*, rev. ed. William Marshall, Friedrich Hempelmann, and Otto zur Strassen (Leipzig, Vienna, 1911), 1:72.

CHAPTER 3

1 Alfred Edmund Brehm, *Das Leben der Vögel* (Glogau: Flemming, 1861), 698.

2 See Herbert Wendt, *Out of Noah's Ark: The Story of Man's Discovery of the Animal Kingdom*, trans. Michael Bullock (Boston: Houghton Mifflin, 1959), 225.

3 Errol Fuller, *Extinct Birds* (New York: Facts on File, 1988), 100.

4 Dieter Luther, *Die ausgestorbenen Vögel der Welt* (Magdeburg: Westarp-Wissenschaften, 1995), 82*ff*.

CHAPTER 4

1 Jean de Léry, *Histoire d'un voyage faict en la terre du Brésil* (2nd. ed. of 1580 edited by Frank Lestringant).

2 Janet Whatley, trans., *History of a Voyage to the Land of Brazil* (Berkeley: University of California Press, 1990), 87; Léry, *Histoire d'un voyage*, 279.

3 The parrot the Tupinamba call the "Canindé" is the Blue-and-yellow Macaw. Because *Ara ararauna* is 30–33 inches long, *Anodorhynchus purpurascens* cannot be confused with the conspicuously longer Hyacinth Macaw, as many ornithologists would claim.

4 Ibid., 88; Léry, 1994, 279.

5 Claude Lévi-Strauss, *Tristes Tropiques*, trans. John and Doreen Weightman (London: Cape, 1973), 219.

6 Claude Lévi-Strauss, "The Raw and the Cooked," trans. John and Doreen Weightman (New York: Harper & Row, 1969), 47.

7 Claude Lévi-Strauss, *Saudades do Brasil: A Photographic Memoir*, trans. Sylvia Modelski (Seattle: University of Washington Press, 1995), 16–17.

8 The Glaucous Macaw is 26–28 inches long.

9 See Tony Juniper, *Spix's Macaw: The Race to Save the World's Rarest Bird* (London: Harper Collins, 2002), 64–65.

10 See Franz Robiller, *Papageien*, 3: 17.

11 See Juniper, *Spix's Macaw*, 84–88.

12 See Juniper, *Spix's Macaw*, 86. Keyhole surgery (laparoscopy) is a harmless procedure per se; the main risk is the anesthetic. See Robiller, *Papageien*, 3:139*ff*.

13 *Anodorhynchus leari* is 30–32 inches long. The bare skin bordering the lower end of the bill is bright yellow and shaped like a half-moon, but it is corn yellow and sickle-shaped in the Hyacinth Macaw.

14 Edward Lear, *Illustrations of the Family of Psittacidae, or Parrots* (London: E. Lear, 1830–32).

15 Quoted in Juniper, *Spix's Macaw*, 73.

16 Caatinga is a type of vegetation charac-
teristic of northeastern Brazil, an arid,
deciduous forest that is green only during
the seasonal rains. Its trees are of low or
medium height, and its thorny brush and
shrubs somehow survive in the driest soil.

17 Johann Baptist von Spix and Carl
Friedrich Philipp von Martius. *Reise
in Brasilien in den Jahren 1817–1820*
(Stuttgart: Brockhaus, 1966), 2:756.

18 Carl Friedrich Philipp von Martius,
Flora Brasiliensis, quoted in Albert
Bettex, *The Discovery of Nature* (New
York: Simon & Schuster, 1965), 201.

19 Spix, *Reise in Brasilien*, 2:528.

20 Johann Baptist von Spix, *Avium species
novae, quas in itinere per Brasiliam Annis
MDCCCXVII–MDCCCXX iussu et auspiciis
Maximiliani Josephi I Bavariae Regis
suscepto collegit et descripsit* (Leipzig:
Fleischer, 1824–25), 1:25.

21 His tombstone in the Alter Südfried-
hof (Old South Cemetery) in Munich
memorializes his bones as those *viri
sagacissimi, integerrimi* (of a most
sagacious man of utmost integrity).

22 Félix de Azara, *Voyages dans l'Amérique
méridionale* [1781–1801] (Paris: Dentu,
1809), 53.

23 CITES is the Convention on International Trade in Endangered Species of
Wild Fauna and Flora, also known as the
Washington Convention. It came into
effect in 1975.

24 Juniper, *Spix's Macaw*, 154.

25 Ibid., 159.

26 Spix, *Reise in Brasilien*, 2:723.

27 Ibid., 2:717.

28 Also called *Tabebuia aurea* because of
the golden trumpet flowers that this tree
of the *Bignoniaceae* family displays from
August to September. In Portuguese it is
called *caraibeira*; common English names
are Silver Trumpet and Tree of Gold, or,
misleadingly, Caribbean Trumpet.
Following Martius's species designation,
the term *Caraiba tree* will be used.

29 See Juniper, *Spix's Macaw*, 144, and
Birdlife International, *Threatened Birds
of the World*, 258.

30 The International Council for Bird
Preservation (ICBP) was founded in 1922
and is the oldest international conservation organization in the world.

31 See Juniper, *Spix's Macaw*, 108.

32 The bird catcher on his bicycle was
transporting three Turquoise-fronted
Amazons (*Amazona aestiva*) and one
Yellow-faced Amazon (*Amazona
xanthops*).

33 The Permanent Committee for the
Recovery of Spix's Macaw was founded
in March 1990. See Juniper, *Spix's
Macaw*, 282.

34 The aggressive Africanized "killer"
honey bees have occupied 40 percent of
macaw nesting holes; it has happened
that swarms of intruding bees have
attacked breeding birds and stung them
to death.

35 See Robiller, *Papageien*, 3:32.

36 Mauricio dos Santos is the trucking
business operator at whose place the
beautiful Spix's Macaw landed after it
was stolen on the Riacho Melància.

37 Juniper, *Spix's Macaw*, 217.

CHAPTER 5

1 Gen. 2:19.

2 [The German word *Waldrapp* is sometimes used in English instead of the
more frequent scientific name Northern
Bald Ibis, occasionally Hermit Ibis. It
is less awkward and is preferred in this
chapter. The word comes from *Wald*,
"forest, wood(s)," and *rapp*, a folk form of
Rabe, "raven, crow" (*Corvus*).—TRANS.]

3 See Marcus zum Lamm, *Die Vogelbücher
aus dem Thesaurus picturarum* (Stuttgart:
Ulmer, 2000), 85.

4 Frederick II of Hohenstaufen, *Art of
Falconry: Being the* De arte venandi cum
avibus *by Frederick II of Hohenstaufen*,
trans. Casey A. Wood and F. Marjorie
Fyfe (Stanford: Stanford University
Press, 1943), 58.

5 Ernst Kantorowicz, *Frederick the
Second: 1194–1250*, trans. E. O. Lorimer
(London: Constable, 1931), 358–59.

6 Pierre Belon du Mans, *L'Histoire de la
nature des oyseaux, avec leurs descriptions, & naïfs portraicts retirez du naturel:
escrite en sept livres* (Geneva: Droz, 1997),
199–200.

7 Conrad Gesner, *Gesneri Redivivi,
aucti & emendati Tomus II. (und III.
= Zweyter Theil) Oder Vollkommenes
Vogel-Buch* (Franckfurt am Mayn: Ilßner,
1669; reprint, Hannover: Schlütersche
Verlagsanstalt, 1995), 2:313.

8 Ibid.

9 Gesner, *Gesneri redivivi*, 3:34–35; trans.
Walter Rothschild, Ernst Hartert, and
Otto Kleinschmidt, "*Comatibis eremita*,"
371–72.

10 Anita Albus, *Paradies und Paradox:
Wunderwerke aus fünf Jahrhunderten*
(Frankfurt am Main: Eichborn, 2002),
219. [The quotation is a self-description
taken from Linnaeus.—TRANS.]

11 See Karin Pegoraro, *Der Waldrapp*, 113.

12 [The German term for "screwing" is
literally "to bird," *vögeln*.—TRANS.]

13 Pegoraro, *Der Waldrapp: Vom Ibis, den
man für einen Raben hielt* (Wiesbaden:
Aula, 1996), 64.

14 Ibid., 58.

15 [There is a double entendre in the
German text here: *schlüpfen* means
both "to slip (out)" and "to hatch
(out)."—TRANS.]

16 Pegoraro, *Der Waldrapp*, 79.

17 See Markus Unsöld, *Sonnenbadever-
halten bei Ibissen* (Threskiornithidae)
(Diploma Thesis, University of Munich,
Zoological Institute, Munich, 2001), 28.

18 [A play on the German word *besonnen*,
"circumspect, prudent," but it is also
meant here to suggest, more literally
and less correctly, "shone upon by the
sun."—TRANS.]

19 See Unsöld, *Sonnenbadeverhalten bei
Ibissen*, 18.

20 Ibid., 78ff.

21 Charles G. Danford, "A further
contribution to the ornithology of
Asia Minor," 81–99.

22 Hans Kumerloeve, "The Waldrapp,
Geronticus eremita (Linnaeus 1758),
und Glattnackenrapp, *Geronticus calvus*
(Boddaert 1783): Zur Geschichte ihrer
Erforschung und zur gegenwärtigen
Bestandssituation," *Annalen des*

Naturhistorischen Museums in Wein 81
(1978): 331. [A brief survey of this topic
in English is in Hans Kumerloeve, "The
Waldrapp, *Geronticus eremite* (Linnaeus,
1758): Historical Review, Taxonomic
History, and Present Status," *Biologi-*
cal Conservation 30 (1984): 363–73.
—TRANS.]

23 See Ragnar K. Kinzelbach and
Walter Nagel, eds., *Ökologie,*
Naturschutz, Umweltschutz, 92.

24 [*Hirsch* also means "stag" in German.
The play on words invokes a German
saying similar in meaning to the English
expression, "to send the fox to guard
the chickens."—TRANS.]

25 Kumerloeve, "Waldrapp," 332–33.

26 Pegoraro, *Der Waldrapp,* 121.

27 Ibid., 120.

28 See Karin Pegoraro et al., "First evidence
of mtDNA sequence differences between
Northern Bald Ibises (*Geronticus*
eremita) of Moroccan and Turkish
origin," *Journal für Ornithologie* 142
(2001): 425–28.

29 See Ellen Thaler et al., "Comeback
des Waldrapp? Ein Pilotversuch zur
Auswilderungsmethodik," Nationalpark
79, no. 2, 26–29.

30 See *International Advisory Group for*
Northern Bald Ibis Newsletter 2 (May
2003).

31 See Johannes Fritz and Angelika Reiter,
eds., *Der Flug des Ibis: Wie ein heiliger*
Vogel durch Menschen fliegen lernt; Die
Rückkehr eines heiligen Vogels aus der
Arche Noah (Vienna: Bibliothek der
Provinz), 2003.

CHAPTER 6

1 [The German word for Corncrake
is *Wachtelkönig*, literally "quail
king."—TRANS.]

2 Johann Andreas Naumann, *Naturge-*
schichte der Vögel Mitteleuropas,
newly revised by G. Berg, R. Blasius,
et al., edited by Carl R. Hennicke
(Gera-Untermhaus: Köhler, 1897–1905),
7:185.

3 Ibid.

4 August Strindberg, *En blå bok: Del 1,*
in *Samlade skrifter av August Strindberg*
(Stockholm: Bonniers, 1918), 46:15.

5 Ibid., 46: 305–6.

6 Johann Georg Gmelin, *D. Johann Georg*
Gmelins Reise durch Sibirien, von dem
Jahr 1733 bis 1743 (Goettingen: Abraham Vandenhoecks Witwe, 1751–52);
new edition published as *Expedition ins*
unbekannte Sibirien, Dittmar Dahlmann,
ed. (Sigmaringen: Thorbecke, 1999),
3:393.

7 Norbert Schäffer, "Habitatwahl und
Partnerschaftssystem von Tüpfelralle
Porzana porzana und Wachtelkönig *Crex*
crex," *Ökologie der Vögel* 21 (1999): 101.

8 Brehm, *Brehms Tierleben: Vögel,* 2:172.

9 Naumann, *Naturgeschichte der Vögel*
Mitteleuropas, 7:186.

10 [Another pun: *Kamm-Musik* sounds
almost like *Kammermusik,* "chamber
music," which is why the bird is called a
Musikant, an amateur music maker, in
the next sentence.—TRANS.]

11 See Schäffer, "Habitatwahl und Partnerschaftssystem," 110–11, and Roal'd
Leonidovich Potapov and Vladimir

Evgenevich Flint, eds., *Handbuch der Vögel der Sowjetunion*, 4:265.

12 Oskar Heinroth and Magdalena Heinroth, *Die Vögel Mitteleuropas: In allen Lebens- und Entwicklungsstufen photographisch aufgenommen und in ihrem Seelenleben bei der Aufzucht vom Ei ab beobachtet*, the Staatliche Stelle für Naturdenkmalpflege in Preußen, ed. (Berlin: Bermühler, 1926–33), 3:77*ff*.

13 Norbert Schäffer et al., "Das Lautrepertoire des Wachtelkönigs," *Vogelwelt* 118 (1997): 147–52.

14 Johann Matthæus Bechstein, *Gemeinnützige Naturgeschichte der Vögel Deutschlands*, 2nd ed. (Leipzig: Crusius, 1809) 3:473.

15 Schäffer et al., "Das Lautrepertoire des Wachtelkönigs," 148.

16 Naumann, *Naturgeschichte der Vögel Mitteleuropas*, 7:188, and Brehm, *Brehms Tierleben: Vögel*, 2:173.

17 Potapov et al., *Handbuch der Vögel der Sowjetunion*, vol. 4: *Galliformes, Gruiformes* (Wittenberg-Lutherstadt: Ziemsen, 1989) 4:274.

18 Brehm, *Brehms Tierleben-Vögel*, 2:173

19 Schäffer, "Habitatwahl und Partnerschaftssystem," 150.

20 Schäffer, "Das Lautrepertoire des Wachtelkönigs," 149*ff*.

21 Quoted in Erwin Stresemann, *Ornithology: From Aristotle to the Present*, ed. G. William Cottrell, trans. Hans J. and Cathleen Epstein (Cambridge, MA: Harvard University Press, 1975), 307. Johann Friedrich Naumann (1780–1857) is the son of Johann Andreas Naumann (1744–1826). Coenraad Jacob

Temminck's *Manuel* was the standard work of European ornithology for a long time; the first two volumes of the second edition appeared in 1820, two more in 1835 and 1840.

22 Moor grass exhibits only one to three knots above the blade's bulb-shaped base; otherwise the blade has no knots throughout its entire length.

23 See Uwe Riecken et al., *Rote Liste der gefährdeten Biotoptypen Deutschlands. Zweite fortgeschriebene Fassung 2006* (Bonn: Bundesamt für Naturschutz, 2007).

24 See Hansjörg Küster, *Geschichte der Landschaft in Mitteleuropa: Von der Eiszeit bis zur Gegenwart* (Munich: Beck, 1995), 69*ff*.

25 Martin Flade, "Wo lebte der Wachtelkönig *Crex crex* in der Urlandschaft?" *Vogelvelt* 118 (1967): 141–46.

26 See Norbert Schäffer, "Der Wachtelkönig: Ein Unbekannter rückt ins Licht," *Der Falke* 43(1996): 317.

CHAPTER 7

1 Pliny the Elder, *The Historie of the World: Commonly called The Natural Historie of C. Plinius Secundus*, trans. Philemon Holland (London: Islip, 1601), 56, 155.

2 [*Nachtschatten*, a vernacular name for the Nightjar; the usual German word is *Nachtschwalbe*, "night swallow."—TRANS.]

3 Aristotle, *History of Animals*, ed. and trans. D.M. Balme, IX (Cambridge, MA: Harvard University Press, 1991), 30.

4 Urs N. Glutz von Blotzheim, Kurt M. Bauer, et al., *Handbuch der Vögel Mitteleuropas* (Frankfurt am Main and Wiesbaden: Akademische Verlagsgesellschaft [through vol. 9]; Wiesbaden: Aula, 1966–97), 9:648.

5 See Claude Lévi-Strauss, *The Jealous Potter*, trans. Bénédicte Chorier (Chicago: University of Chicago Press, 1988), 34–37.

6 Oskar Heinroth, "Beobachtungen bei der Zucht des Ziegenmelkers (*Caprimulgus europaeus* L.)," *Journal für Ornithologie* 57 (1909): 67.

7 Ibid.

8 See Heinroth, *Die Vögel Mitteleuropas*, 1:277.

9 Heinroth, "Beobachtungen," 69.

10 Glutz von Blotzheim, 9:659.

11 Audubon observed that the feathery bristles of the Whip-poor-will, *Caprimulgus vociferus*, primarily aid in the folding of the wings of large moths, which are always caught from behind. See Audubon, *Ornithological Biography*, 1:425.

12 See Reiner Schlegel, *Der Ziegenmelker: Caprimulgus europaeus* L. (Magdeburg: Westarp, 1995), 8, 18; and Glutz von Blotzheim, *Handbuch der Vögel Mitteleuropas*, 9:658.

13 Albert Krambrich, "Aus dem Leben der Nachtschwalbe," *Vogelwelt* 75 (1954): 57.

14 Alfred Edmund Brehm and Emil Adolf Roßmäßler, *Die Thiere des Waldes* (Leipzig: Winter, 1864–67), 361.

15 Naumann, *Naturgeschichte der Vögel Mitteleuropas*, 4:249.

16 See Glutz von Blotzheim, *Handbuch der Vögel Mitteleuropas*, 9:659.

17 See Schlegel, *Der Ziegenmelker*, 28–29, and Glutz von Blotzheim, 9:648.

18 *Rueet* corresponds to Brehm's *heh-eet*. Heinroth hears it as *qvik*.

19 See Schlegel, *Der Ziegenmelker*, 29.

20 Ibid., 30.

21 Brehm, *Brehms Tierleben*, 3:279.

22 See Heinroth, "Beobachtungen," 68.

23 Ibid., 62.

24 Heinroth, *Die Vögel Mitteleuropas*, 1:279.

25 Heinroth, "Beobachtungen," 63.

26 Ibid.

27 See Schlegel, *Der Ziegenmelker*, 38–39.

28 Heinroth, "Beobachtungen," 74.

29 Ibid., 64.

30 Ibid., 65–66.

31 See Schlegel, *Der Ziegenmelker*, 57–58.

32 Ibid., 54.

33 Heinroth, "Beobachtungen," 70–71.

34 Ibid., 67–68.

35 Schlegel, *Der Ziegenmelker*, 52.

36 Ibid., 58.

37 Ibid., 75.

38 See Heinroth, "Beobachtungen," 71.

39 Ibid., 70.

40 See Schlegel, *Der Ziegenmelker*, 66.

41 The Whip-poor-will is called *Caprimulgus vociferous*, the *Klagenachtschwalbe*, "crying nightjar," or the *Schwarzkehl-Nachtschwalbe*, "black-throated nightjar." The smaller bird that Thomas Nuttall discovered, and that was given the name *Phalaenoptilus nuttallii* in his honor, was first called Nuttall's Whip-poor-will; "Common Poorwill" became established later.

42 Translated by John and Doreen Weightman as *The Jealous Potter*, 14–16.

43 Oral communication from Josef. H. Reichholf, to whom I also owe the reference to the Plenter forest problem.

CHAPTER 8

1 Lorenz Oken, *Allgemeine Naturgeschichte für alle Stände, von Professor Oken* (Stuttgart: Hoffman, 1837), 7:125.

2 See Wolfgang Epple and Manfred Rogl, *Schleiereulen Geister der Nacht—Faszination und Aberglaube* (Karlsruhe: Braun, 1993), 34.

3 Ibid., 38.

4 Ibid., 40.

5 Ibid., 25.

6 Ibid., 44.

7 Ibid., 64.

8 Ibid., 51–52.

9 Ibid., 60.

10 See Josef H. Reichholf, "Sind Hauskatzen Nahrungskonkurrenten der Schleiereulen *(Tyto alba)*?" *Eulen-Rundblick* 51–52 (2004): 11–14.

CHAPTER 9

1 The subspecies *Surnia ulula ulula* in the northern Paleoarctic is distinct from two other subspecies, *Surnia ulula tianschanica* in the Central Asian mountains and *Surnia ulula caparoch* in northern North America.

2 See Theodor Mebs, *Die Eulen Europas: Biologie, Kennzeichen, Bestände* (Stuttgart: Kosmos, 2000), 366, and Wolfgang Scherzinger, "Sperbereulen—Außenseiter aus der Taiga," *Gefiederte Welt* 5 (2001): 174.

3 Brehm, *Brehms Tierleben*, 3:220.

4 Glutz von Blotzheim, *Handbuch der Vögel Mitteleuropas*, 363–64. [David Sibley's description of the sounds made by North American Hawk Owls is this: "Courtship call (heard mainly at night) a series of popping whistles up to six seconds long *popopopopo*... Female and juvenile give weak, screeching *tshoooIP.*' Also a thin, rising whistle *feeeee*. Alarm a shrill, chirping *quiquiquiqui*." David Sibley, *The Sibley Guide to Birds* (New York: Knopf, 2000), 283.—TRANS.]

5 See Mebs, *Die Eulen Europas*, 165.

6 Mainly the little vole, *Clethrionomys glareolus*, less frequently lemmings.

7 For the mythological significance of the Hawk Owl in Russia, see Albus, *Paradies und Paradox*, 273–86.

CHAPTER 10

1 Having copulated.

2 Ovid, *Ovid's Metamorphoses, in Fifteen Books: Translated by the Most Eminent Hands; Adorned with Sculptures*, XI, 1060–69, trans. John Dryden et al., ed. Sir Samuel Garth (London: Jacob Tonson, 1717), 405.

3 See Aristotle, *History of Animals*, IX, 14, 616a, 19–33, trans. D.M. Balme, 277–79.

4 Pliny the Elder, *The Historie of the World*, x, xlvii, 90–91, trans. Philemon Holland, 179.

5 Plutarch, "Whether Land or Sea Animals Are Cleverer," XII, 35, 983, *Plutarch's Moralia*, trans. Harold

Cherniss and William C. Helmbold
(Cambridge, MA: Harvard University
Press, 1968), 461, 463.

6 Belon, *L'Histoire de la nature des oyseaux*,
218–19.

7 Georg Agricola, *Zwölf Bücher vom
Berg- und Hüttenwesen*, ed. and trans.
Carl Schiffner (Munich: Deutscher
Taschenbuch Verlag, 1994), 521.

8 The first stanza of "Der Kirche Trutz
und Schutz (To Defend and Protect
the Church)," by Sigmund von Birken;
quoted in Wolfgang Harms, "Der
Eisvogel und die hakyonischen Tage," in
Verbum et Signum, 1: 477–515, eds. Hans
Fromm, Wolfgang Harms, and Uwe
Ruberg (Munich: Fink, 1975), 500.

9 See Jacob and Wilhelm Grimm,
Deutsches Wörterbuch [German
Dictionary], under *Eisvogel*.

10 [For an explanation of how structural
color is based on light waves and not
pigments, see Anita Albus, *The Art of
Arts*, 99–106.—TRANS.]

11 Naumann, *Naturgeschichte der Vögel
Mitteleuropas*, 4:347.

12 Brehm, *Das Leben der Vögel*, 496.

13 Naumann, *Naturgeschichte der Vögel
Mitteleuropas*, 4:349–50.

14 Heinroth, *Die Vögel Mitteleuropas*,
1:286.

15 Ibid., 1:287.

16 See Margret Bunzel-Drüke and Joachim
Drüke, *Eisvögel* (Karlsruhe: Braun,
1996), 52.

17 Ibid., 55.

18 This is why digging up a Kingfisher
burrow is strictly prohibited.

19 Bunzel-Drüke, *Eisvögel*, 64.

20 Ibid., 45.

21 *Ispida* comes from *ispidus*, 'bristly,'
'prickly.'

22 Glutz von Blotzheim, *Handbuch*,
9:743–44.

23 Naumann, *Naturgeschichte der Vögel
Mitteleuropas*, 4:360–61.

24 Glutz von Blotzheim, *Handbuch*,
9:783ff.

25 Audubon, *Ornithological Biography*,
1:395.

26 I am grateful to Claude Lévi-Strauss for
drawing my attention to this fable and
other non-European Kingfisher legends
and myths.

27 Erminnie Smith, *Myths of the Iroquois*
(Ohsweken, ON: IPACS, 1983), 75–77.

28 Franz Boas, *The Mythology of the Bella
Coola Indians*, Memoirs of the American Museum of Natural History 2,
pt. 2 (New York: American Museum
of Natural History, 1898), 33–34, and
pl. IX fig. 8.

AFTERWORD

1 "BUT YOU, BEAST, / The more I regard
you / the more I become MAN / In Mind,
O Beast." "Paraboles: pour accompagner
douze aquarelles de L. Albert-Lasard,"
Œuvres, 1:20. Translated by Hilary
Corke in Paul Valéry, *Poems in the Rough*
(Princeton: Princeton University Press,
1969), 6.

2 BirdLife International, *Threatened Birds
of the World*. [The figures that follow are
from 2009.]

3 Brehm, *Brehms Tierleben: Vögel*, 3:7.

4 Ragnar K. Kinzelbach, and Walter Nagel, eds., *Ökologie, Naturschutz, Umweltschutz*, Dimensionen der modernen Biologie 6 (Darmstadt: Wissenschaftliche Buchgesellschaft, 1989), 84.

5 Irenäus Eibl-Eibesfeldt, *Grundriß der vergleichenden Verhaltungsforschung. Ethologie*, 7th ed. (Munich: Piper, 1987), 619.

6 Georges Louis Leclerc, Comte de Buffon, quoted in D'Arcy Wentworth Thompson, *Growth and Form* (Cambridge: Cambridge University Press, 1942), 1020.

7 Eibl-Eibesfeldt, ibid.

8 "Rudolph Wagner... [m]y good & kind agent for the propagation of the Gospel ie the Devil's gospel." Charles Darwin to T. H. Huxley, 8 August 1860, in *Correspondence of Charles Darwin* (Cambridge: Cambridge University Press, 1993), 8:316. [The reference is specifically to *Origin of Species* but is probably meant to apply broadly.—TRANS.]

9 Edward O. Wilson, *The Diversity of Life* (Cambridge, MA: Belknap Press, 1992), 347–50.

10 Ibid., 272.

11 Ibid., 351; see also 384.

12 Claude Lévi-Strauss, *View from Afar*, trans. Joachim Neugroschel and Phoebe Hoss (New York: Basic Books, 1985), 31–32.

13 Wolf Singer, "Verschaltungen legen uns fest," 30–65.

14 Erwin Panofsky, *Meaning in the Visual Arts* (Garden City, NY: Doubleday, 1955), 3.

15 Wolf Singer, *Ein neues Menschenbild? Gespräche über Hirnforschung* (Frankfurt am Main: Suhrkamp, 2003), 21. [For a brief statement in English of this leading neuroscientist's views relating to these points, see Wolfgang Singer, "Understanding the brain: How can our intuition fail so fundamentally when it comes to studying the organ to which it owes its existence?" *EMBO reports*, special issue, 8 (2007): 16–19. —TRANS.]

16 Giovanni Pico della Mirandola, *Oration of the Dignity of Man*, trans. Robert Caponigri (Chicago: Gateway Editions, 1956), 8, 7.

17 Robert Spaemann, *Philosophische Essays* (Stuttgart: Reclam, 1994), 57.

18 Jakob von Uexküll, *Umwelt und Innenwelt der Tiere* (Berlin: Springer, 1921), 4.

19 Ernst Cassirer, *Ziele und Wege der Wirklichkeitserkenntnis*, eds. John Michael Krois and Klaus Christian Köhnke (Hamburg: Meiner, 1999), 85–86.

20 I thank Josef H. Reichholf for communicating this true event to me. To prevent any confusion with parrots living or dead, I have changed the parrot's name...

21 Claude Lévi-Strauss, *The Savage Mind*, trans. John and Doreen Weightman (Chicago: University of Chicago Press, 1966), 204.

22 Claude Lévi-Strauss, *The Naked Man*, trans. John and Doreen Weightman (New York: Harper & Row, 1981), 691.

23 See Petra Busch, *Die Vogelparlamente und Voglesprachen in der deutschen*

Literatur des Späten Mittelalters und der Frühen Neuzeit (Munich: Fink, 2001).

24 A bird's feathers, hollow bones, and air bags are light, *"léger,"* just like the poet's actual surname, *Léger.*

25 Saint-John Perse, "Birds," trans. Robert Fitzgerald, *Collected Poems* (Princeton: Princeton University Press, 1971), 625, 627.

26 Plato, *Timaeus* 91 E, trans. R. G. Bury, in *Plato, With an English Translation* (Cambridge, MA: Harvard University Press, 1929), 7:251.

27 Jacobus de Voragine, *Golden Legend of Jacobus de Voragine,* trans. Granger Ryan and Helumt Ripperger (New York: Arno Press, 1969), 606.

28 Quoted in Knut Hagberg, *Carl Linnaeus,* trans. Alan Blair (London: Cape, 1952), 149–50.

29 Eibl-Eibesfeldt, *Grundriß* 249.

30 Edgar Wind, *Experiment and Metaphysics,* trans. Cyril Edwards (Oxford: Legenda, 2001), 130.

31 Lévi-Strauss, *Naked Man,* 679–90, esp. 682.

32 Jacques Roger, *Buffon,* trans. Sarah Lucille Bonnefoi (Ithaca, NY: Cornell University Press, 1997), 252.

33 Ernst Mayr, *Growth of Biological Thought* (Cambridge, MA: Belknap Press, 1982), 623–24.

34 Ernst Cassirer, *Essay on Man* (New Haven: Yale University Press, 1944), 83, and Claude Lévi-Strauss, *Mythos und Bedeutung,* ed. Adalbert Reif (Frankfurt am Main: Suhrkamp, 1980), 236–51.

35 Saint-John Perse, "Birds," 639.

"DISCOURSE ON THE NATURE OF BIRDS"

1 Anita Albus wishes to thank Karin Pegoraro and Manfred Föger for their assistance in preparing the endnotes. [This essay, "Discours sur la nature des oiseaux," is the introductory essay from Buffon's *Histoire naturelle des Oiseaux,* 16 vols. (Paris, 1770–86), 1:3–60. The translation by William Smellie is a somewhat abridged version from 1793, published in London. His influential version remained the most popular one throughout the next century. The original spelling has been retained in Smellie's text but not in the endnotes; Buffon's footnotes were also not included in Albus's German translation.—TRANS.]

2 The sense of touch as the guarantor of knowledge is foreign to us today. Buffon wrote in the volume of his *Histoire naturelle, générale et particulière* devoted to humans: "It is only through the sense of touch that we can gain complete and true knowledge; it is the one sense that corrects all others which would produce but illusions and errors in our mind, were it not that the sense of touch teaches us to form judgments." [Smellie: "It is by the sense of feeling alone that we acquire real knowledge. The innumerable errors into which we are led by the illusions of the other senses are corrected by feeling." vol. 3, sect. 8, p. 48.]

3 Mammals travel great distances as well, e.g., gnus and reindeer, but especially bats and whales, though bats and whales cannot be considered mammals.

4 Large birds such as storks, geese, and ibises migrate with their parents, but in the case of small birds, the adults often migrate weeks before their young do.

5 But if a bird lacks "nostrils" it will always have "nasal slits."

6 Buffon writes that these three senses literally "give each creature their dominant sensations." Today we consider *Homo sapiens*, along with birds and higher primates, as "visually oriented animals."

7 Toads and frogs were still counted among the reptiles in Buffon's day. [Smellie's "flight buzz of insects" should be "calls of reptiles," which explains Albus's note. Her translation is very accurate and congenial.—TRANS.]

8 This error appears as well in the beautiful "In Praise of Birds" by Giacomo Leopardi: "Some say, and it would be relevant here, that the voices of birds are more sweet and appealing, and their song more modulated, in our parts of the world, than in those where men are savage and crude; and they conclude that birds, even though free, acquire some small part of the civilization of the men whose dwelling-places they frequent." *Moral Tales: Operette Morali*, trans. Patrick Creagh, *Works of Giacomo Leopardi* (Manchester: Carcanet New Press, 1983), 1:165; originally published as "Elogio degli uccelli [1824]," in *Operette morali* (Naples: Guida, 1977). [A note in Leopardi's manuscript of the essay refers to this same section in Buffon.—TRANS.]

9 Appearances are deceiving. The most intelligent birds are parrots and corvids (the crow family).

10 The male Nightingale does not help construct the nest and does not provide the female with food. [The references are to parts of Albus's full translation of Buffon's text.—TRANS.]

11 Birds actually do live longer. But it's very different as far as sexual maturity is concerned. Small birds like tits (or chickadees) and finches are able to reproduce at around eight months, whereas the Bearded Vulture is only sexually mature at six years. Buffon does not even take into account the difference between birds that stay in the nest (altricial birds) and those that leave it early (precocial birds).

12 The vulture eats carrion, buzzards (hawks in the buteo family) eat primarily small mammals, and the beaver is a vegetarian.

13 There are insectivores among the quadrupeds, too: anteaters, bat-eared foxes, etc.

14 Buffon is wrong about this.

15 Buffon is often in the dark about what birds eat.

16 The bird's most important sensation of touch is located around its beak (sensitive tip, whiskers at the base).

17 This is unfortunately true of only some species.

18 Today we know that migratory birds have a highly developed long-term memory. Scientists have discovered that these birds have a markedly expanded hippocampus, where spatial information is processed in the brain.

Bibliography

Agricola, Georg. *Zwölf Bücher vom Berg- und Hüttenwesen: in denen die Ämter, Instru-*
mente, Maschinen und alle Dinge, die zum Berg- und Hüttenwesen gehören, nicht nur aufs
deutlichste beschrieben, sondern auch durch Abbildungen, die am gehörigen Ort eingefügt
sind, unter Angabe der lateinischen und deutschen Bezeichnungen aufs klarste vor Augen
gestellt werden; Sowie sein Buch von den Lebewesen unter Tage. Edited and translated by
Carl Schiffner, with the assistance of Ernst Darmstaedter et al. Munich: Deutscher
Taschenbuch Verlag, 1994. Originally published as *De re metallica* (Basel, 1556).

Albus, Anita. *The Art of Arts: Rediscovering Painting.* Translated by Michael Robertson.
New York: Knopf, 2000. Originally published as *Die Kunst der Künste: Erinnerungen*
an die Malerei (Frankfurt am Main, 1997).

——. *Paradies und Paradox: Wunderwerke aus fünf Jahrhunderten.* Frankfurt am Main:
Eichborn, 2002.

Aristotle. *History of Animals.* Translated by D. M. Balme. Cambridge, MA: Harvard
University Press, 1991.

Audubon, John James. *Ornithological Biography, or An Account of the Habits of the Birds of*
the United States of America; Accompanied by Descriptions of the Objects Represented in
the Work entitled The Birds Of America, and Interspersed with Delineations of American
Scenery and Manners. 5 vols. Philadelphia: Dobson, 1831–39.

Azara, Félix de. *Voyages dans l'Amérique méridionale* [1781–1801]. Paris: Dentu, 1809.

Barrère, Pierre. *Ornithologiae Specimen Novum, sive Series Avium in Ruscinone, Pyrenaeis*
Montibus, atque in Galliâ Æquinoctiali Observatarum, in Classes, genera & species,
novâ methodo, digesta. Perpignan: Le Comte, 1745. Facsimile, Geneva: Droz, 1996.

Bechstein, Johann Matthæus. *Gemeinnützige Naturgeschichte der Vögel Deutschlands.*
2nd ed. Vol. 3. Leipzig: Crusius, 1809.

Belon du Mans, Pierre. *L'Histoire de la nature des oyseaux, avec leurs descriptions, & naïfs portraicts retirez du naturel: escrite en sept livres.* Paris: Guillaume Cauellat, 1555. Facsimile. (Introduced and annotated by Philippe Glardon) Geneva: Droz, 1997.

Bettex, Albert. *The Discovery of Nature.* New York: Simon and Schuster, 1965. Originally published as *Die Entdeckung der Natur* (Munich, 1965).

Birdlife International. *Threatened Birds of the World.* Cambridge / Barcelona: Birdlife International / Lynx Ediciones, 2000.

Boas, Franz. *The Mythology of the Bella Coola Indians.* Memoirs of the American Museum of Natural History 2, pt. 2. New York: American Museum of Natural History, 1898.

Brehm, Alfred Edmund. *Brehms Tierleben: Vögel.* 4 vols. Revised edition by William Marshall, Friedrich Hempelmann, and Otto zur Strassen. Leipzig: Bibliographisches Institut, 1911–14.

———. *Das Leben der Vögel: Dargestellt für Haus und Familie.* Glogau: Flemming, 1861.

Brehm, Alfred Edmund, and Emil Adolf Roßmäßler. *Die Thiere des Waldes.* 2 vols. Leipzig: Winter, 1864–67.

Brisson, Mathurin Jacques. *Ornithologia, sive synopsis methodica sistens avium divisionem in ordines, sectiones, genera, species, ipsarumque varietates.* 6 vols. Paris: Bauche, 1760–63.

Buffon, Georges Louis Leclerc, Comte de. "On the Nature of Birds." Translated by William Smellie. In *The Natural History of Birds: From the French of the Count de Buffon. Illustrated with engravings; and a preface, notes, and additions, by the Translator.* 1: 1—350. London: A. Strahan, T. Cadell, and J. Murray, 1793. Originally published as "Discours sur la nature des oiseaux." In *Histoire naturelle des oiseaux.* Vol. 1 (Paris, 1770).

Bunzel-Drüke, Margret, and Joachim Drüke. *Eisvögel: Faszinierende Meisterfischer in bedrohten Lebensräumen.* Karlsruhe: Braun, 1996.

Busch, Petra. *Die Vogelparlamente und Vogelsprachen in der deutschen Literatur des Späten Mittelalters und der Frühen Neuzeit.* Beihefte zu Poetica, vol. 24. Munich: Fink, 2001.

Cassirer, Ernst. *An Essay on Man: An Introduction to a Philosophy of Human Culture.* New Haven: Yale University Press, 1944.

———. *Ziele und Wege der Wirklichkeitserkenntnis.* Edited by John Michael Krois and Klaus Christian Köhnke. Nachgelassene Manuskripte und Texte 2. Hamburg: Meiner, 1999.

Cokinos, Christopher. *Hope Is the Thing with Feathers: A Personal Chronicle of Vanished Birds.* New York: Tarcher/Putnam, 2000.

Danford, Charles G. "A further contribution to the ornithology of Asia Minor." *The Ibis,* 4, no. 4 (1880): 81–99.

Darwin, Charles. *The Correspondence of Charles Darwin.* Vol. 8: 1860. Edited by Frederick Burkhardt et al. Cambridge: Cambridge University Press, 1993.

Eibl-Eibesfeldt, Irenäus. *Grundriß der vergleichenden Verhaltensforschung: Ethologie.* 7th ed. Munich: Piper, 1987.

Epple, Wolfgang, and Manfred Rogl. *Schleiereulen Geister der Nacht—Faszination und Aberglaube.* Karlsruhe: Braun, 1993.

Flade, Martin. "Wo lebte der Wachtelkönig *Crex crex* in der Urlandschaft?" *Vogelwelt* 118 (1967): 141–46.

Frederick II of Hohenstaufen. *The Art of Falconry: Being the* De arte venandi cum avibus *by Frederick II of Hohenstaufen.* Translated and edited by Casey A. Wood and F. Marjorie Fyfe. Stanford: Stanford University Press, 1943. Originally published as *De arte venandi cum avibus* (Ausburg, 1596); written in 1247.

Fritz, Johannes, and Angelika Reiter, eds. *Der Flug des Ibis: Wie ein heiliger Vogel durch Menschen fliegen lernt; Die Rückkehr eines heiligen Vogels aus der Arche Noah.* Vienna: Bibliothek der Provinz, 2003.

Fuller, Errol. *Extinct Birds.* New York: Facts on File, 1988.

Gesner, Conrad. *Gesneri redivivi, aucti & emendati Tomus II. (und III. = Zweyter Theil) Oder Vollkommenes Vogel-Buch: Darstellend Eine wahrhafftige und nach dem Leben vorgerissene Abbildung Aller / so wol in den Lüfften und Klüfften / als in den Wäldern und Feldern / und sonsten auff den Wassern und daheim in den Häusern / nicht nur in Europa / sondern auch in Asia, Africa, America, und anderen neu-erfundenen Ost- und West-Indischen Insulen / sich enthaltender zahmer und wilder Vögel und Feder-Viehes.* Franckfurt am Mayn: Ilßner, 1669. Reprint (with an afterword by Werner Steinigeweg), Hannover: Schlütersche Verlagsanstalt, 1995.

Glutz von Blotzheim, Urs N., Kurt M. Bauer, et al., eds. *Handbuch der Vögel Mitteleuropas,* 14 vols. in 22 parts. Frankfurt am Main and Wiesbaden: Akademische Verlagsgesellschaft [through vol. 9]; Wiesbaden: Aula, 1966–97.

Gmelin, Johann Georg. *D. Johann Georg Gmelins Reise durch Sibirien, von dem Jahr 1733 bis 1743.* 4 vols. Goettingen: Abraham Vandenhoecks Witwe, 1751–52. New edition published as *Expedition ins unbekannte Sibirien.* Edited by Dittmar Dahlmann. Sigmaringen: Thorbecke, 1999.

Hagberg, Knut. *Carl Linnaeus.* Translated by Alan Blair. London: Cape, 1952. Originally published as *Carl Linnæus* (Stockholm, 1939).

Harms, Wolfgang. "Der Eisvogel und die halkyonischen Tage: Zum Verhältnis von naturkundlicher Beschreibung und allegorischer Naturdeutung." In *Verbum et Signum: Beiträge zur mediävistischen Bedeutungsforschung [Festschrift für Friedrich Ohly],* 1: 477–515. Edited by Hans Fromm, Wolfgang Harms, and Uwe Ruberg. Munich: Fink, 1975.

Heinroth, Oskar. "Beobachtungen bei der Zucht des Ziegenmelkers (*Caprimulgus europaeus* L.)." *Journal für Ornithologie* 57 (1909): 56–83.

Heinroth, Oskar, and Magdalena Heinroth. *Die Vögel Mitteleuropas: In allen Lebens- und Entwicklungsstufen photographisch aufgenommen und in ihrem Seelenleben bei der Aufzucht vom Ei ab beobachtet.* Edited by the Staatliche Stelle für Naturdenkmalpflege in Preußen. 4 vols. Berlin: Bermühler, 1926–33.

Humboldt, Alexander von. "Über die Wasserfälle des Orinoco [1808]." In *Ansichten der Natur mit wissenschaftlichen Erläuterungen.* 3rd expanded ed. Stuttgart: Cotta, 1849.

———. *Views of Nature: Or Contemplations of the Sublime Phenomena of Creation; with Scientific Illustrations.* Translated by E. C. Otté and Henry G. Bohn. London: Bohn, 1850.

International Advisory Group for Northern Bald Ibis Newsletter 2 (May 2003).

Juniper, Tony. *Spix's Macaw: The Race to Save the World's Rarest Bird.* London: Harper Collins, 2002.

Kantorowicz, Ernst H. *Frederick the Second: 1194–1250.* Translated by E. O. Lorimer. London: Constable, 1931. Originally published as *Kaiser Friedrich der Zweite* (Berlin, 1927).

Kinzelbach, Ragnar K., and Walter Nagel, eds. *Ökologie, Naturschutz, Umweltschutz.* Dimensionen der modernen Biologie 6. Darmstadt: Wissenschaftliche Buchgesellschaft, 1989.

Krambrich, Albert. "Aus dem Leben der Nachtschwalbe." *Vogelwelt* 75 (1954): 100–101.

Kumerloeve, Hans. "The Waldrapp, *Geronticus eremita* (Linnaeus, 1758): Historical Review, Taxonomic History, and Present Status." *Biological Conservation* 30 (1984): 363–73.

———. "Waldrapp, *Geronticus eremita* (Linnaeus 1758), und Glattnackenrapp, *Geronticus calvus* (Boddaert 1783): Zur Geschichte ihrer Erforschung und zur gegenwärtigen Bestandssituation." *Annalen des Naturhistorischen Museums in Wien* 81 (1978): 319–49.

Küster, Hansjörg. *Geschichte der Landschaft in Mitteleuropa: Von der Eiszeit bis zur Gegenwart.* Munich: Beck, 1995.

Lamm, Marcus zum. *Die Vogelbücher aus dem Thesaurus Picturarum.* Edited, with commentary, by Ragnar K. Kinzelbach and Jochen Hölzinger. Stuttgart: Ulmer, 2000.

Lear, Edward. *Illustrations of the Family of Psittacidae, or Parrots: The Greater Part of them Species Hitherto Unfigured; Containing Forty-two Lithographic Plates, Drawn from Life, and on Stone.* London: E. Lear, 1830–32.

Leopardi, Giacomo. *Moral Tales: Operette Morali.* Translated by Patrick Creagh. *Works of Giacomo Leopardi.* Vol. 1. (Manchester: Carcanet New Press, 1983). Originally published as *Operette morali* (Naples, 1977); first published in 1827.

Léry, Jean de. *History of a Voyage to the Land of Brazil, otherwise called America: Containing the Navigation and the Remarkable Things Seen on the Sea by the Author; the Behavior of Villegagnon in That Country; the Customs and Strange Ways of Life of the American Savages; Together with the Description of Various Animals, Trees, Plants, and Other Singular Things Completely Unknown over Here.* Edited and translated by Janet Whatley. Berkeley: University of California Press, 1990. Originally published as *Histoire d'un voyage faict en la terre du Brésil* (Geneva, 1578).

Lévi-Strauss, Claude. *The Jealous Potter.* Translated by Bénédicte Chorier. Chicago: University of Chicago Press, 1988. Originally published as *La Potière jalouse* (Paris, 1985).

———. *Mythos und Bedeutung: Fünf Radiovorträge; Gespräche mit Claude Lévi-Strauss.* Edited by Adalbert Reif. Frankfurt am Main: Suhrkamp, 1980.

———. *The Naked Man.* Introduction to a Science of Mythology 4. Translated by John and Doreen Weightman. New York: Harper & Row, 1981. Originally published as *L'Homme nu* (Paris, 1971).

———. *The Raw and the Cooked.* Vol. 1, Introduction to a Science of Mythology. Translated by John and Doreen Weightman. New York: Harper & Row, 1969. Originally published as *Le cru et le cuit* (Paris, 1950).

———. *Saudades do Brasil: A Photographic Memoir.* Translated by Sylvia Modelski. Seattle: University of Washington Press, 1995. Originally published as *Saudades do Brasil* (Paris, 1994).

———. *The Savage Mind.* Translated by John and Doreen Weightman. Chicago: University of Chicago Press, 1966. Originally published as *La Pensée sauvage* (Paris, 1962).

———. *Tristes Tropiques.* Translated by John and Doreen Weightman. London: Cape, 1973. Originally published as *Tristes Tropiques* (Paris, 1955).

———. *The View from Afar.* Translated by Joachim Neugroschel and Phoebe Hoss. New York: Basic Books, 1985. Originally published as *Le Regard éloigné* (Paris, 1983).

Luther, Dieter. *Die ausgestorbenen Vögel der Welt.* Die Neue Brehm-Bücherei 424. Magdeburg: Westarp-Wissenschaften, 1995.

Martius, Carl Friedrich Philipp von. *Flora Brasiliensis, enumeratio plantarum in Brasilia hactenus detectarum.* Vol. 1. Munich: Fleischer, 1840.

Mayr, Ernst. *The Growth of Biological Thought: Diversity, Evolution, and Inheritance.* Cambridge, MA: Belknap Press, 1982.

Mebs, Theodor, and Wolfgang Scherzinger. *Die Eulen Europas: Biologie, Kennzeichen, Bestände.* Stuttgart: Kosmos, 2000.

Naumann, Johann Andreas. *Naturgeschichte der Vögel Mitteleuropas.* Newly revised by G. Berg, R. Blasius, et al. Edited by Carl R. Hennicke. 12 vols. Gera-Untermhaus: Köhler, 1897–1905.

Nissen, Claus. *Die illustrierten Vogelbücher.* Stuttgart: Hiersemann, 1953.

Oken, Lorenz. *Allgemeine Naturgeschichte für alle Stände von Professor Oken.* Bearbeitet von F. A. Walchner. 8 vols. Stuttgart: Hoffman, 1833–42.

Ovid. *Ovid's Metamorphoses, in Fifteen Books: Translated by the Most Eminent Hands; Adorned with Sculptures.* Edited by Sir Samuel Garth. Translated by John Dryden et al. London: Jacob Tonson, 1717.

Panofsky, Erwin. *Meaning in the Visual Arts: Papers in and on Art History.* Garden City, NY: Doubleday, 1955.

Pegoraro, Karin. *Der Waldrapp: Vom Ibis, den man für einen Raben hielt.* Wiesbaden: Aula, 1996.

Pegoraro, Karin, Manfred Föger, and Walther Parson. "First evidence of mtDNA sequence differences between Northern Bald Ibises (*Geronticus eremita*) of Moroccan and Turkish origin." *Journal für Ornithologie* 142, no. 4 (October 2001): 425–28.

Pico della Mirandola, Giovanni. *Oration on the Dignity of Man.* Translated by Robert Caponigri. Chicago: Gateway Editions, 1956. Originally published as *De hominis dignitate* (Bologna, 1496).

Plato. *Timaeus.* In *Plato, With an English Translation.* Translated by R. G. Bury. 7:16–253. Cambridge, MA: Harvard University Press, 1929.

Pliny the Elder. *The Historie of the World: Commonly called, The Naturall Historie of C. Plinius Secundus. Historia naturalis.* Translated by Philemon Holland. London: Islip, 1601.

Plutarch. "Whether Land or Sea Animals Are Cleverer." In *Plutarch's Moralia: In Fifteen Volumes.* Vol. XII, 920 A–999 B. Translated by Harold Cherniss and William C. Helmbold. Cambridge, MA: Harvard University Press, 1968.

Potapov, Roal'd Leonidovich, and Vladimir Evgenevich Flint, eds. *Handbuch der Vögel der Sowjetunion.* Vol. 4: *Galliformes, Gruiformes.* Wittenberg-Lutherstadt: Ziemsen, 1989.

Quammen, David. *The Song of the Dodo: Island Biogeography in an Age of Extinctions.* New York: Scribner, 1996.

Reichholf, Josef H. "Sind Hauskatzen Nahrungskonkurrenten der Schleiereulen (*Tyto alba*)?" *Eulen-Rundblick* 51–52 (2004): 11–14.

Riecken, Uwe, et al. *Rote Liste der gefährdeten Biotopen Deutschlands. Zweite fortgeschriebene Fassung 2006.* Bonn: Bundesamt für Naturschutz, 2007.

Robiller, Franz. *Papageien.* Vol. 3: *Mittel- und Südamerika.* Stuttgart: Ulmer, 1990.

Roger, Jacques. *Buffon: A Life in Natural History.* Edited by L. Pearce Williams. Translated by Sarah Lucille Bonnefoi. Ithaca, NY: Cornell University Press, 1997. Originally published as *Buffon: Un philosophe au Jardin du Roi* (Paris, 1989).

Rothschild, Walter, Ernst Hartert, and Otto Kleinschmidt. "*Comatibis eremita* (LINN.), a European bird." *Novitates Zoologicae* 4, no. 3 (December 1897): 371–80.

Saint-John Perse (Alexis Saint-Léger Léger). *Collected Poems.* Translated by W. H. Auden et al. Bollingen Series 87. Princeton: Princeton University Press, 1971.

Schäffer, Norbert. "Habitatwahl und Partnerschaftssystem von Tüpfelralle *Porzana porzana* und Wachtelkönig *Crex crex*." *Ökologie der Vögel* 21 (1999): 1–267.

————. "Der Wachtelkönig: Ein Unbekannter rückt ins Licht." *Der Falke* 43 (November 1996): 316–19.

Schäffer, Norbert, Urte Salzer, and Dieter Wend. "Das Lautrepertoire des Wachtelkönigs *Crex crex*." *Vogelwelt* 118 (1997): 147–56.

Scherzinger, Wolfgang. "Sperbereulen—Außenseiter aus der Taiga." *Gefiederte Welt* 5 (2001): 173ff.

Schlegel, Reiner. *Der Ziegenmelker*: Caprimulgus europaeus L. Die Neue Brehm-Bücherei 406. Wittenberg-Lutherstadt: Ziemens, 1969. Reprint, Magdeburg: Westarp-Wissenschaften, 1995.

Sibley, David. *The Sibley Guide to Birds*. New York: Knopf, 2000.

Singer, Wolf. *Ein neues Menschenbild? Gespräche über Hirnforschung*. Frankfurt am Main: Suhrkamp, 2003.

————. "Understanding the brain: How can our intuition fail so fundamentally when it comes to studying the organ to which it owes its existence?" *EMBO* reports, special issue, 8 (2007): 16–19.

————. "Verschaltungen legen uns fest: Wir sollten aufhören, von Freiheit zu sprechen." In *Hirnforschung und Willensfreiheit: Zur Deutung der neuesten Experimente*. Edited by Christian Geyer. pp. 30–65, 2387. Frankfurt am Main: Suhrkamp, 2004.

Smith, Erminnie A. *Myths of the Iroquois*. Ohsweken, ON: IPACS, 1983. Originally published in the *Second Annual Report of the Bureau of Ethnology to the Secretary of the Smithsonian Institution 1880–1881* (Washington, 1883).

Spaemann, Robert. *Philosophische Essays*. Expanded ed. Stuttgart: Reclam, 1994.

Spix, Johann Baptist von. *Avium species novae, quas in itinere per Brasiliam Annis MDCCCXVII–MDCCCXX iussu et auspiciis Maximiliani Josephi I. Bavariae Regis suscepto collegit et descripsit*. 2 vols. Leipzig: Fleischer, 1824–25.

Spix, Johann Baptist von, and Carl Friedrich Philipp Martius. *Reise in Brasilien in den Jahren 1817–1820*. Edited by Karl Mägdefrau. 4 vols. Stuttgart: Brockhaus, 1966. Reprint of the 1823–31 edition.

Stresemann, Erwin. *Ornithology: From Aristotle to the Present*. Edited by G. William Cottrell. Translated by Hans J. and Cathleen Epstein. Cambridge, MA: Harvard University Press, 1975. Originally published as *Die Entwicklung der Ornithologie von Aristoteles bis zur Gegenwart* (Berlin, 1951).

Strindberg, August. *En blå bok: Del 1* [A blue book: Part 1]. In *Samlade skrifter av August Strindberg*. Vol. 46. Stockholm: Bonniers, 1918.

Thaler, Ellen, Karin Pegoraro, and Susanne Stabinger. "Comeback des Waldrapp? Ein Pilotversuch zur Auswilderungsmethodik." *Nationalpark* 79, no. 2 (1993): 26–29.

Thompson, D'Arcy Wentworth. *On Growth and Form*. New ed. Cambridge: Cambridge University Press, 1942.

Turner, William. *Avium praecipuarum, quarum apud Plinium et Aristotelem mentio est, brevis et succincta historia.* Cologne: Johann Gymnich, 1544.

Uexküll, Jakob von. *Umwelt und Innenwelt der Tiere.* Berlin: Springer, 1921.

Unsöld, Markus. *Sonnenbadeverhalten bei Ibissen* (Threskiornithidae). Diploma Thesis. University of Munich, Zoological Institute. Munich, 2001.

Valéry, Paul. *Œuvres.* Edited by Jean Hytier. Bibliothèque de la Pléiade. 2 vols. Paris: Gallimard, 1957–60.

———. *Poems in the Rough.* Translated by Hilary Corke. Bollingen Series 45/2. Princeton: Princeton University Press, 1969.

Voragine, Jacobus de. *The Golden Legend of Jacobus de Voragine.* Translated and adapted by Granger Ryan and Helmut Ripperger. New York: Arno Press, 1969. Originally published as *Legenda aurea,* compiled circa 1260.

Wendt, Herbert. *Out of Noah's Ark: The Story of Man's Discovery of the Animal Kingdom.* Translated by Michael Bullock. Boston: Houghton Mifflin, 1959. Originally published as *Auf Noahs Spuren: Die Entdeckung der Tiere* (Berlin, 1956).

Wilson, Alexander. *Wilson's American Ornithology, with notes by Jardine: To Which Is Added a Synopsis of American Birds, Including Those Described by Bonaparte, Audubon, Nuttall, and Richardson.* Edited by T. M. Brewer. Boston: Otis Broaders & Co., 1840. www.wilsonsociety.org/AWilsonplates/parakeettext.html (accessed March 4, 2009).

Wilson, Edward O. *The Diversity of Life.* Cambridge, MA: Belknap Press, 1992.

Wind, Edgar. *Experiment and Metaphysics: Towards a Resolution of the Cosmological Antinomies.* Translated by Cyril Edwards. Oxford: Legenda, 2001. Originally published as *Das Experiment und die Metaphysik: Zur Auflösung der kosmologischen Antinomien* (Tübingen, 1934).

List of Illustrations and Credits

FREQUENTLY CITED WORKS AND ILLUSTRATORS

Albus   Anita Albus.

Audubon   John James Audubon. *The Birds of America, from Original Drawings by John James Audubon*. 4 vols. London: Published by the author, 1827–38.

Belon   Pierre Belon du Mans. *L'Histoire de la nature des oyseaux*. Paris: Guillaume Cavellat, 1555. Facsimile. Geneva: Droz, 1997.

Buhle   Charles Adam Adolph Buhle. *Die Naturgeschichte in getreuen Abbildungen und mit ausführlicher Beschreibung derselben*. Vol. 2, *Vögel*. Leipzig: Eisenach, 1835.

Gesner   Conrad Gesner. *Historiae Animalium Liber III, qui est de auium natura. Adjecti sunt ab initio indices alphabetici decem super nominibus avium in totidem linguis diversis*. Francofurti: Henricii Laurentii, 1617.

Heinroth   Oskar Heinroth. *Die Vögel Mitteleuropas*. 4 vols. Berlin: Bermühler, 1926–33. No owner of the rights to these books could be ascertained. The S. Fischer Verlag, Frankfurt am Main, is committed to pay royalties according to the usual rates for any legal claims.

Keulemans   John Gerrard Keulemans. In Johann Andreas Naumann. *Naturgeschichte der Vögel Mitteleuropas*. 12 vols. Gera: Köhler, 1897–1905.

Lear   Edward Lear. *Illustrations of the Family of Psittacidae, or Parrots*. London: Published by the author, 1832.

Naumann   Johann Andreas Naumann. *Naturgeschichte der Vögel Mitteleuropas*. Newly revised by G. Berg, R. Blasius, et al. Edited by Carl R. Hennicke. 12 vols. Gera: Köhler, 1897–1905.

155 Heinroth, vol. 2, *Barn Owls*, 1926–33, colored photographs: 1. *at 6 days*; 2. *nestling at 35 days, in second downy plumage*; 3. *adult.*

156 Heinroth, vol. 2, *Barn Owl No. 1*, 1926–33, photographs: 1. *at 18 hours, in first downy plumage*; 2. *at 27 days, in second downy plumage*; 3. *at 35 days*; 4. *at 42 days*; 5. *and* 6. *at 47 days, dropping feathers*; 7. *at 54 days, able to fly*; 8. *at 61 days*; 9. *and* 10. *at 79 days, fully grown*; 9. *active, ready to fly*; 10. *drawn up in alarm.*

157 Heinroth, vol. 2, *Barn Owl No. 2*, 1926–33, photographs: 1. *at 16 days*; 2. *at 27 days*; 3. *at 42 days*; 4. *to* 6. *adult, active*; 7. *somewhat alarmed*; 8. *more alarmed*; 9. *dead Barn Owl with right side of head bare, showing large front ear covering (tagus) behind the eye.*

161 Albus, *Barn Owl "Head over Heels,"* watercolor and body color on paper, n.d.

163 Albus, *Barn Owl with Pellets, Alarmed by Day in Attic*, oil on copper, n.d.

170 Belon, *Barn Owl*, 1555.

174 Albus, *Hawk Owl (Surnia ulula), in Birch Forest*, oil on copper, n.d.

175 Albus, *Dorsal View of Hawk Owl Skin*, watercolor and body color on paper, n.d.

178 Aububon, *Hawk Owls*, pl. CCCLXXVIII, 1827–38.

179 Keulemans, *Pair of Hawk Owls*, 1897–1905.

181 Buhle, *Hawk Owl*, 1835.

184 Johann Theodor Susemihl, *The Common Kingfisher (Alcedo ispida)*, gouache, ca. 1800; reproduced with the kind permission of the Hessisches Landesmuseum, Darmstadt.

187 Belon, *Kingfisher (Halcyonium)*, 1555.

189 Johann Ulrich Krauß, *Kingfisher Emblem*, in *Les Tapisseries du Roy, ou sont representez les quatre elemens et les quatre saisons*, Augsburg: Kopfmayer, 1690.

190 Albus, *Pair of Kingfishers (Alcedo ispida) in a Landscape*, watercolor and body color on vellum, n.d.

192 Gesner, *Kingfisher*, 1617.

194 Albus, *Kingfisher with Trout*, still life (detail), watercolor and body color on vellum, n.d.

197 Heinroth, vol. 1, *Kingfisher (Alcedo ispida)*, 1926–33, colored photographs: 1. *naked nestling with egg*; 2. *9 days before fledging*; 3. *adult.*

199 Keulemans, *Kingfishers (Alcedo ispida)*, 1897–1905.

203 Albus, *Still Life with White-throated Kingfisher Skin (Halcyon smyrnensis)*, gouache on vellum, n.d.

205 Keulemans, *White-throated Kingfishers (Halcyon smyrnensis)*, 1897–1905.

206 John Gould, *Laughing Kookaburras (Dacelo novaeguineae)*, lithograph, in John Gould *The Birds of Australia*, 7 vols. London: published by the author, 1840–48.

209 Audubon, *Belted Kingfishers (Megaceryle alcyon)*, pl. LXXVII, 1827–38.

210 Albus, *Kingfisher (Alcedo ispida) in a Miniature*, watercolor and body color on vellum, n.d.

227 Belon, *Human and Avian Skeletons Compared*, 1555.

Index of Names

Names in small caps indicate
individual birds.
Page numbers in italics indicate
image captions.

Adam, 53, 63
Adanson, Michel, 239
Aeolus, 183
Agricola, Georg, 188
Aharoni, Israel, 80–81
Albertus Magnus, 63
Albin, Eleazar, 75, 78
Albus, Anita, ix, *56, 89, 91, 109, 149, 154,
  161, 163, 174, 175*, 190, 194, 203, 210,
  252n10, 256n7 (chap. 9), 257n10,
  259n1, 260n7
Alcyone (Halcyone), 183, 187, 196
Aldrovandi, Ulisse, 66, 72, 73
Alighieri, Dante, 63–64
ALU, Waldrapp, 103
Aristotle, 68, 105, 125, 183, 185
Astley, Reverend, 37
Audubon, John James, 1–2, 4, *9, 10, 11, 12,
  13*, 152, 208, 209, 249n10, 255n11
Azara, Don Félix de, 41

Barraband, Jacques, 16
Barrère, Pierre, 75
Bechstein, Mattæus Johann, 79, 117
Belon du Mans, Pierre, 66, *67, 68*, 73,
  106, 170, 187, 188, 225, 227
Bettex, Albert, 251n18
Birken, Sigmund von, 257n8
BLACK, Waldrapp, 103
BLUE, Waldrapp, 103
Boas, Franz, 257n28
Bonaparte, Charles Lucien, 37, 41
Brehm, Alfred Edmund, 21, 107, 108,
  110, 111, 118, 132, 134, 136, 137,
  176, 192, 216, 250n6 (chap. 2),
  255n18
Brisson, Mathurin-Jacques, 75–76,
Buffon, Georges-Louis Leclerc, Comte de,
  ix, 77, 110, 226–28, 229–48, 259n1,
  259n2, 260n6, 260n7, 260n8, 260n10,
  260n11, 260n14, 260n15
Buhle, Christian Adam Adolf, 3, 43, 65,
  74, *181*
Bunzel-Drüke, Margaret and Joachim
  Drüke, 257n16
Busch, Petra, 258n23

273